Atmospheric Turbulence

Atmospheric Turbulence: A Molecular Dynamics Perspective

Adrian F. Tuck

OXFORD
UNIVERSITY PRESS

OXFORD
UNIVERSITY PRESS

Great Clarendon Street, Oxford OX2 6DP

Oxford University Press is a department of the University of Oxford.
It furthers the University's objective of excellence in research, scholarship,
and education by publishing worldwide in

Oxford New York

Auckland Cape Town Dar es Salaam Hong Kong Karachi
Kuala Lumpur Madrid Melbourne Mexico City Nairobi
New Delhi Shanghai Taipei Toronto

With offices in

Argentina Austria Brazil Chile Czech Republic France Greece
Guatemala Hungary Italy Japan Poland Portugal Singapore
South Korea Switzerland Thailand Turkey Ukraine Vietnam

Oxford is a registered trade mark of Oxford University Press
in the UK and in certain other countries

Published in the United States
by Oxford University Press Inc., New York

First published 2008

British Library Cataloguing in Publication Data
Data available

Library of Congress Cataloging in Publication Data
Data available

Typeset by Newgen Imaging Systems (P) Ltd., Chennai, India
Printed in Great Britain
on acid-free paper by
Biddles Ltd., King's Lynn, Norfolk

ISBN 978–0–19–923653–4

10 9 8 7 6 5 4 3 2 1

Dedication

The author dedicates this book to his mentors P. G. Ashmore, B. A. Thrush, K. R. Wilson, and R. J. Murgatroyd, who in their different ways were everything scientists should be.

Preface

This book grew out of a series of four invited lectures given at the Meteorology Department, University of Reading, in February 2006, at the instigation of Professor Alan O'Neill, director of the NERC Data Assimilation Research Centre. The questions asked there served to sharpen some of the points which had been presented. The material in those lectures drew on work published in the literature during the previous seven years, applying Schertzer and Lovejoy's theory of generalized scale invariance to a large body of airborne data which clearly showed atmospheric variability well above instrumental noise levels but which could not be described adequately by Gaussian PDFs and second moment power spectra.

The atmosphere is molecules in motion, but a lacuna exists as regards explicit discussion or treatment of this fact in the meteorological literature and among standard texts on dynamic meteorology, fluid mechanics, turbulence, multifractals, non-equilibrium statistical mechanics, and kinetic molecular theory. While texts on atmospheric chemistry of course deal in molecular behaviour, the step from kinetic molecular theory to atmospheric motion is made often without comment, usually via application of the law of mass action on scales many orders of magnitude larger than the mean free path and on which true diffusion cannot be dominant. Discussions of the progression from molecular to fluid motion are found mainly in the statistical mechanics literature but with no consideration of complicated anisotropic large-scale flows within morphologically irregular boundaries, such as those the atmosphere exhibits. There are few examples of attempts to examine the molecular roots of turbulence. This book aims to point out the need to address this situation, and to offer suggestions about how to proceed.

The central point of this book is that molecular dynamics, via the generation of vorticity in the presence of anisotropies such as gravity, planetary rotation, and the solar beam, influences the structure of turbulence, temperature, radiative transfer, and chemistry in the atmosphere. Because energy is deposited in the atmosphere by molecules absorbing photons, energy must propagate upward from the smallest scales. Analyses are presented of observations by the statistical multifractal methods developed by Schertzer and Lovejoy, which show generalized scale invariance in the atmosphere. The need to unite molecular dynamics, turbulence theory, fluid mechanics, and non-equilibrium statistical mechanics is reinforced by the fact that core wind speeds in jet streams can exceed one third of the most probable velocity of air molecules, a breach of the conditions under which standard derivations of the Navier–Stokes equation are made. Note that in saying this I do not intend to imply that continuum fluid mechanics needs major reformulation in the context of the meteorological simulation of the large-scale flow by numerical process on computers for weather forecasting; the enterprise is too demonstrably successful for that to be the case. Indeed, some understanding of the success of this operation emerges naturally from analysing high-resolution observations in a statistical multifractal framework. However, for representing the smaller scales, and for accurate accounting of the detailed energy distribution in the atmosphere, required for climate prediction, turbulence must be properly understood and formulated. It is my contention that it will not be achieved without explicit recognition of the fact that fluid mechanical behaviour emerges spontaneously in the molecular dynamics simulation of a population of Maxwellian molecules subject to an anisotropy (Alder and Wainwright 1970): turbulence has molecular roots.

A word of justification is necessary for the heavy emphasis on aircraft data in this book, and on the use of dropsondes deployed from aircraft in approaching the vertical behaviour of atmospheric scaling. These sorts of observations reached, in the late 1980s, a level of accuracy, response, sensitivity, and continuity over a large enough range of scales that fairly reliable calculation of probability distributions and the moments involved in statistical multifractal analysis, especially of the rather robust H exponent, was possible. In a smaller number of cases, the scaling exponents characterizing the intermittency and the Lévy distribution also could be calculated. It is still true, however, that there is only a very small number of platforms which have produced observations of the requisite quality. We can look forward to the time when satellite data can sustain such analysis, when numerical models of the atmosphere can simulate it, and when unmanned, autonomous aircraft can circle the globe to extend it. We are not there yet, however, and in the meantime the reader's indulgence must be sought if this book seems a little like an account of the ER-2's history of deployments for stratospheric ozone research. The fact is that the platform-instrument combination achieved a level of performance in acquiring in situ data that has not yet been matched in our context, even though it was far from perfect. I defend what has been done because the implications may be far-reaching, both conceptually and practically.

In my experience, few meteorologists have much deep acquaintance with molecules and their behaviour, while at the same time most physical chemists have limited appreciation of continuum fluid mechanics, particularly in its unique and complicated atmospheric form. The two meet in the persons of those who seek to include chemistry in global atmospheric computer models, but often in such a way that the scales important to our argument here are missing. While it is almost a cliché to say that chemistry, radiation, and dynamics are coupled to produce the atmospheric state, the point is that all three of these are coupled at very short time and space scales in such a way that the concept of local thermodynamic equilibrium is undermined. The long tail in the PDF of molecular speeds is kept overpopulated by its mutually sustaining interaction with vortices induced by the fast molecules themselves in reacting to anisotropy, such as that inherent in gravity, the solar beam, planetary rotation, the surface topography, or larger-scale winds, particularly the fast moving cores of jet streams. Translationally hot fragments from photodissociating ozone molecules therefore do not recoil into a Maxwellian bath of equilibrated air molecules, but into pre-existing vorticity structures. The resulting sustained overpopulation of fast molecules will thus affect the shapes of the pressure-broadened spectral lines of water vapour and carbon dioxide so central in climate, the rates of chemical reactions central to the ozone and aerosol distributions, the phase changes of water, and also the very interpretation of what atmospheric temperature itself means. An attempt has been made where possible, in the spirit of Berry, Rice, and Ross's *Physical Chemistry*, to explain from basic principles what is involved in each argument before dealing with the complexities. However, that book sets a high standard to which to adhere and a size of over a thousand pages is not justified for the present work. References have been provided for the reader willing to branch out into unfamiliar territory.

In order to keep the text compact, the equations have been stated without derivation or proof, giving references and bibliographic sources instead. For the same reason, subjects such as acoustics, aerodynamics, or shock waves have not been considered. The arguments if correct will have considerable implications for the phase transitions of water underlying cloud physics and aerosol behaviour, but only the possibility that long-tailed velocity PDFs of water molecules will affect the interaction of vapour phase molecules with liquid droplets and ice crystals has been pointed out.

The use of the phrase 'atmospheric turbulence' in the title is deliberate, in that it emphasizes the point that true laminar flow is never observed in the atmosphere—speed and directional wind shears are ubiquitous. Thus the phrase should be interpreted in the context of this book as meaning 'the ubiquitous, omnipresent, scale invariant, fluctuating structure of atmospheric quantities'. It is not limited to the episodes intermittently encountered by aircraft that are severe enough to discommode the aircrew and passengers, if any.

During the writing of this book, I came to appreciate Peter Hobbs's remark in the Preface to *Ice Physics*, that authors do not finish books, they merely abandon them.

It remains to acknowledge my debt to the many colleagues whose efforts provided the observational and analytical foundations upon which this book is built. Primary among these is Susan Hovde, without whose mathematical and data analysis skills the work in this book would not have been possible. Some other colleagues in the Meteorological Chemistry Program at the erstwhile NOAA Aeronomy Laboratory (since 1st October 2005 the Chemical Sciences Division of the Earth System Research Laboratory) were also foremost in the airborne research: Kenneth Aikin, David Fahey, Ru-Shan Gao, Jeffrey Hicke, Kenneth Kelly, Daniel Murphy, Michael Proffitt, Eric Ray, Stephen Reid, Erik Richard, Karen Rosenlof, Arthur Schmeltekopf, Xin Tao, Thomas Thompson, Ian Watterson, and Richard Winkler. The book benefited from readings by David Fahey and Jamie Donaldson that exceeded the calls of friendship, in addition to the anonymous reviews commissioned by OUP, which led to extensive and material improvements. I also express my appreciation of the enlightened directorship of the NOAA Aeronomy Laboratory by Daniel Albritton during the period 1986–2005. The meteorological observations from the ER-2 were made by T. Paul Bui from NASA Ames Research Center, the chlorine monoxide observations were made by James Anderson from Harvard, and the nitrous oxide observations were made by James Podolske and Max Loewenstein from NASA Ames. The Microwave Temperature Profiler data were taken by Bruce Gary and Michael Mahoney of the Jet Propulsion Laboratory, at the California Institute of Technology. Jeffrey Hicke did the analysis and MM5 modelling for the 19980411 WB57F flight. None of the data would have been taken without the efforts of the aircrew and ground crew of the aircraft: the ER-2 and DC-8 at NASA Ames, Moffett Field, the WB57F at NASA Johnson Space Center, Ellington Field, and the Gulfstream 4SP at the NOAA Air Operations Center, Tampa Bay. Generous funding for the missions using the ER-2, DC-8, and WB57F came from the NASA Upper Atmosphere Research Program. Funds and support from the NASA programmes by Robert Watson, Michael Kurylo, and Donald Anderson were of primary importance. The results of flux calculations using wind and ozone observations from the UK Meteorological Research Flight C130 W Mk. 2 in the early 1980s by Geraint Vaughan and Daniel McKenna sowed seeds upon the nature of variability that lay dormant for over a decade before germinating when seen in a statistical multifractal framework, after initial irrigation by Nile floods as described by Hurst. The author owes a very particular debt to Daniel Schertzer and Shaun Lovejoy, the originators of generalized scale invariance and its application to the atmosphere. Finally, I gained a great deal from the constant encouragement of my wife, Professor Veronica Vaida, who also sustained many original scientific discussions and was the embodiment of research done the right way, for the right reasons.

The faults of the book are my responsibility alone.

Adrian Tuck
Boulder, Colorado
27 April 2007

Contents

1 Introduction

The atmosphere consists of molecules in motion, yet it is often hard to find any mention of the fact in meteorological texts. This absence is also true of substantial areas of physics and chemistry which have evolved to provide quantitative descriptions of the behaviour of atoms and molecules in the gas phase: in particular, non-equilibrium statistical mechanics and molecular dynamics have had less overlap with the theory and observation of turbulence than perhaps might have been expected. Meteorology of course has had fluid mechanics at front and centre for over a century and has had to face issues in turbulence for over half that time. The purpose of this book is to show that atmospheric turbulence is an emergent property arising from the anisotropic environment of populations of gas molecules, linking molecular dynamics with fluid mechanics through the generation of vorticity. The anisotropies arise from gravity, planetary rotation, the solar beam, and the nature of the topography, the sea and ice surfaces, and the vegetative cover. We shall see that analysis of high resolution data of adequate quality, as yet available largely from only a few aircraft, leads to the emergence of a correlation of the multifractal, turbulent scaling at the smaller scales with some characteristics of the larger scale meteorological flow, such as the intensity and depth of jet streams.

Lest the reader should think that the formulation of events at the microscopic scale has little or nothing to do with the central concerns of modern meteorology, we note that climate is determined through the absorption and emission of photons by molecules in the atmosphere and at the surface. The nature and distribution of these molecules is determined by photochemical kinetics acting in the presence of turbulent transport and biogeochemical fluxes from the surface. Quantitative calculation of the rates of these processes must necessarily account for both the skewed probability distributions of molecular velocities maintained interactively by vorticity structures and the effect of the scale invariant turbulent structures, on such large volumes of chemical reaction as the stratospheric polar vortex. One cannot expect to apply the law of mass action to such meteorological features as if isotropic three-dimensional diffusion was the sole transport mechanism and expect quantitative simulation, except by accident.

In defence of the standard assumption that energy is deposited into the atmosphere on the largest scale, so enabling a cascade of energy down to the smallest, dissipative, scales it might be argued that half of the planet is illuminated by the solar beam at any one time. While that is true, the incident solar photon flux is highly non-uniform over the sunlit hemisphere. Moreover, the photons are absorbed and emitted by molecules whose abundance, phase, radiative contribution, and spatial distribution are determined by turbulent vorticity structures and their interaction with the surface; this importantly includes water in its gaseous, liquid, and solid states. Energy input therefore is influenced on all scales; conservative cascades of energy from large to small scales cannot therefore occur. The atmospheric vorticity structures, the ocean surface, the land surface and its vegetative cover have all been observed to exhibit scale invariance; they behave as statistical multifractals. Complete understanding can only emerge by use of mathematical expression of the relationships linking molecular dynamics and non-equilibrium statistical mechanics; it will necessarily include turbulent, fluid mechanical, physico-chemical, and probabilistic concepts. The reader should note that unfamiliar terms and acronyms have been explained in full in the Glossary section at the back of the book.

1.1 History

Historically observation has mainly led theory. An important exception is the advent of numerical simulation by electronic computer of the emergence of fluid behaviour from flux-driven populations of Maxwellian molecules (Alder and Wainwright 1970). As early as 1500 Leonardo da Vinci wrote acute descriptions of fluid flow, even using the word 'turbulenza'. The first attempts at mathematical formulation of fluid flow came in the second half of the eighteenth century, by such mathematicians as the Bernoullis, Euler, Lagrange, d'Alembert, and Laplace, and long pre-dated knowledge of the atomic and molecular nature of gases and liquids. In the first half of the nineteenth century Navier and Stokes derived a differential equation describing the time evolution of fluid flow. Following earlier work by D. Bernoulli and J. Herapath, J. J. Waterston developed an original formulation of what became known as kinetic molecular theory, including the first statement of the law of equipartition of energy among molecules and that temperature was related to the square of molecular velocity, but the work was rejected by referees and remained unknown until its rediscovery in journal files and publication 47 years later by Rayleigh (Waterston 1845, 1892). By that time Maxwell and Boltzmann had published their seminal works on the subject which, vitally, derived expressions for the probability distributions of molecular velocities. Interaction between Kelvin and Stokes in the second half of the nineteenth century produced the mathematical notion of circulation and Helmholtz initiated the concept of local spin of a fluid element about its axis, an idea related shortly thereafter to the thermal structure by Beltrami and which was apparently first called 'vorticity' in the 4th edition of Lamb's *Hydrodynamics* in 1916. In 1888 Reynolds published his observations of fluid flow through a pipe, establishing the transition from laminar to turbulent flow when the dimensionless ratio of the product of the characteristic velocity and length scales to the kinematic viscosity was of order 10^3.

At the end of the nineteenth century and the beginning of the twentieth, the Norwegian school of meteorology, led by V. Bjerknes, built upon the earlier attempts by Ferrel in the USA to formulate models of cyclogenesis underpinned by differential equations. In the second decade of the twentieth century Enskog and Chapman independently extended Boltzmann's and Gibbs's formulations of kinetic theory and statistical thermodynamics to permit calculation of the transport properties of gases, but could only do so for cases involving small, linear perturbations to the molecular velocity distributions. At about the same time Richardson discovered that the rate of increase of separation distance x of two particles in atmospheric flow was neither exponential or diffusive, but that after time t the quantity x^2t^{-1} was proportional to $x^{4/3}$, that is to say it was power law. This, and his report to the Admiralty and the Treasury that the length of the coastline of Britain depended in a power law manner on the scale of the map, eventually led to what is now called fractality. This resolution of a dispute about how many coastguards per mile were necessary to enforce contraband laws probably pleased neither sailors nor economists, even less that the length of the coastline was infinite! Richardson also proposed weather prediction by numerical calculation of the pressure tendency in 1922, using human rather than mechanical or electronic computers.

In the 1936–40 period, Rossby formulated potential vorticity, a scalar conserved in the limits of adiabaticity and no mixing, by normalizing vorticity to the depth of the column of air under consideration; it was effectively the product of the horizontal inertial and vertical convective instability terms. Ertel provided a general mathematical derivation in 1942 and its use was further explored after the Second World War by Kleinschmidt and by Reed and Danielsen, who used it to study trans-tropopause transport by folding, a term first used by Sawyer in connection with jet stream dynamics. During the war, Sutcliffe and Petterssen independently formulated vorticity-based theories of cyclone development, and in the post-war period Charney and Eady had framed somewhat different theories of baroclinic instability. Eady also wrote a remarkable paper on the cause of the general circulation of the atmosphere for the Royal Meteorological Society's centennial celebration in 1950, in which he argued that the turbulent transfer of heat was causal, with cellular circulations and jet streams being secondary phenomena. The advent of the electronic computer encouraged von Neumann to point out that it would enable the production of weather forecasts by numerical integration of the governing equations forward in time from an observed state. It is hard to overstate the importance of this development; von Neumann and Charney collaborated with Fjortoft to produce the first such attempt. Ironically, it took three decades before computer simulation extended to isentropic maps and potential vorticity fields, thereby leading to fresh understanding of atmospheric transport and development.

During the Second World War Kolmogorov embodied Richardson's idea of a spectrum of eddies of different sizes in his celebrated $k^{-5/3}$ law, one of the few exact results in turbulence theory. However, it was criticized as early as 1944 by Landau on the grounds that it did not incorporate the effects of what is now known as intermittency; the $-5/3$ law was derived independently by Onsager in 1945 and separately by Heisenberg and by von Weizsäcker during their internment and interrogation at Hall Farm after the German surrender. In 1949 Batchelor and Townsend published results describing turbulence at very large wavenumbers k as consisting of small scale vorticity structures such as rolls, slabs, and sheets, a result supporting Landau's criticism.

Towards the end of the 1950s through to the early 1970s, Alder and his co-workers invented molecular dynamics, the computer study of a population of Maxwellian molecules subject to some dis-equilibrating flux or force. Crucially, it was shown how long tails in the molecular velocity autocorrelation function led to 'ring currents', or vortices, on very short time and space scales. Hydrodynamic behaviour had emerged from a randomized population of 'billiard ball' molecules subjected to an anisotropy.

Interest in the specific molecules composing the atmosphere has a long history, playing a central role in the realization of the atomic and molecular nature of gases through the work of Lavoisier, Priestley, Dalton, Gay-Lussac, and Avogadro. Tyndall realized the importance of the infra red absorption by water vapour in 1861, with Arrhenius pointing out in 1896 that extra carbon dioxide in the air from burning coal and oil could raise the temperature at the Earth's surface by re-emitting more infrared radiation towards the surface, allowing less to escape to space. Schönbein discovered ozone in the mid-nineteenth century, with its presence in the atmosphere being deduced via its ultraviolet absorption by Hartley in 1881. Fabry and Buisson proposed that ozone was produced by the action of sunlight on molecular oxygen; Cornu and Dobson both deduced that it must be in the upper atmosphere. By 1930 Chapman had formulated a photochemical mechanism involving oxygen only to account for ozone production, writing an account of the quantum mechanical properties of atoms and molecules in his presidential address to the Royal Meteorological Society in 1934, even pointing out that if astronomers wished to make a hole in the ozone layer to observe the ultraviolet spectra of stars they would need to deploy a catalytic agent and wear protective clothing.

Using coarse representations of the absorption and emission of radiation by water vapour, carbon dioxide, and ozone, Simpson had in 1929 deduced that the atmosphere had an excess of radiative energy incident upon it at latitudes less than 35° and emitted a net excess of radiation to space at higher latitudes, arguing that this was what drove the atmospheric circulation. Bates and Nicolet suggested in 1950 that chain reactions carried by H, OH, and HO_2 limited the abundance of ozone in the mesosphere, and by the mid-1960s Hampson had proposed that the reactions of the electronically excited $O(^1D)$ atom observed from ozone photolysis in Norrish's laboratories were effective in producing OH and HO_2 from water vapour, later proposing a similar production of NO and NO_2 from nitrous oxide. The latter idea was also taken up by Crutzen, who realized the role of nitric acid and the cross-coupling of the different chain reactions. Molina and Rowland pointed out the high efficiency of the chain reactions carried by Cl and ClO arising from photodissociation of $CFCl_3$ and CF_2Cl_2 which had lifetimes of decades after use as aerosol can spray propellants and refrigerants. These halogen emissions culminated in the ozone hole, a phenomenon which launched several extensive airborne missions producing in situ observations of meteorological and chemical data good enough to sustain multifractal analysis of the turbulent structure.

Building on the pioneering, original work by Mandelbrot in the 1970s, Schertzer and Lovejoy formulated the turbulent structure of the atmosphere as a statistical multifractal in the 1980s. At the same time, Paltridge revisited Simpson's approach to the radiative forcing of the meridional circulation in terms of the need to maximize entropy production, a perspective being taken up afresh in the light of a recent mathematical derivation of the maximum entropy principle by Dewar, using Jaynes's exposition of statistical mechanics.

A glance at the contemporary atmospheric literature shows global numerical models producing ever more complicated simulations of weather and climate. How well are the basic couplings between radiation, turbulent dynamics, and chemistry handled, given that they occur over 15 orders of magnitude in spatial scale in the atmosphere, from the mean free path near the surface to the length of a great circle?

1.2 The importance of computers

Whether one attempts to solve the Boltzmann H-equation for a population of air molecules or the Navier–Stokes equation for fluid flow in some atmospheric domain, the non-linearity of the equations precludes analytical solutions. Nevertheless, treating the evolving atmospheric state as an initial value problem in computational physics, so arriving at future solutions of the state of the mass of moving, reacting, and radiating molecules, is a prospect which cannot fail to impress, both as regards intellectual content and as regards utility. The success of the endeavour, on time scales of days for specific synoptic scale weather forecasts and on the time scale of the past century for more statistical climate simulation, ranks as one of the major scientific accomplishments in the second half of the twentieth century. Concise reviews are available for both weather forecasting (Bengtsson 1999) and for climate modelling (Mitchell 2004). An invaluable perspective in the context of the need for improved parametrization of smaller scale, unresolvable processes has been provided (Palmer 2001). The need for parametrization may readily be appreciated by noting that the current state of the art for a priori molecular dynamics simulations is about 10^{10} atoms, largely dictated by the difficulty of parallelization of the collisions and trajectories (Kadau et al. 2004). Nevertheless, these authors did succeed in simulating the onset of Rayleigh–Taylor instability at an interface between two fluids. The atmosphere has about 8×10^{43} molecules. If one alternately views it as a top-down rather than as a bottom-up exercise, the number of degrees of freedom of turbulence scales as the 9/4 power of the Reynolds number, $\mathrm{Re}^{9/4}$, resulting in enormous requirements for computer power (Succi 2001; Davidson 2004), given that Re for the free atmosphere is of order 10^{12}. The closure problem for the non-linear Navier–Stokes equation, that the statistical expression for the nth moment of a variable involves its $(n+1)^{\mathrm{th}}$ power, causes a fundamental difficulty. It may perhaps be necessary to attempt parametrization from the small scales up, without sacrificing a statistically adequate description of basic molecular behaviour and the way in which it generates vorticity and influences the absorption and emission of photons. The scale invariance observed in the atmosphere, from an Earth radius down through five orders of magnitude to tens of metres, may offer some useful constraints, particularly if it can also be shown to emerge with the correct scaling exponents from molecular dynamical simulation. The range of scales remains daunting however, with eight decades dividing the largest currently possible molecular dynamics scales and the smallest currently possible observational ones.

1.3 Airborne observations

This book largely uses airborne observations, with some from sondes dropped on parachutes from research aircraft. Given the diversity of observations by many techniques at the surface, the routine launches of radiosondes worldwide, and the many successful observations of pressure, temperature, humidity, and many chemical species by remote sounding from orbit, it is reasonable to ask why it is necessary to be so selective. The basic reason is that once scale invariance had been observed to be a property of atmospheric variables, requirements emerged as regards noise levels, continuity, range, scale, and platform characteristics which were too restrictive to be met by the majority of observations. For example, air observations at a fixed point on the surface have no spatial range and are mainly made at hourly, semidiurnal, or diurnal frequencies; the sensors on ascending radiosonde balloons are in the turbulent wake of the balloon; satellite instruments do not yet simultaneously have the vertical and horizontal resolution to provide useable statistics on the smaller scales. The wind field is particularly demanding to observe remotely. These limitations could change in the future, but the requirement to resolve at least three decades in spatial scale makes it difficult to achieve from a satellite moving at about $7\,\mathrm{km\,s^{-1}}$. Satellites and long-range autonomous aircraft are the only means to extend the analysis of scale invariance to the global scale. To extend it to small scales, a few metres, is currently possible from aircraft and dropsondes, but it will be challenging to design instruments capable of fast enough time response at good enough signal-to-noise to span the gap from metres to the tens of nanometres required to examine fluctuations induced by molecular motion responding to anisotropies. The absence of data gaps is essential if the intermittency and Lévy exponents, which characterize the multifractality of the observations, are to be calculated, an important limitation for many data sources. These exponents are defined later, in Chapter 4. The indications from the research aircraft observations are, however, that the problem will need to be tackled for a complete description of atmospheric motion to be achieved. Just as they have in treating the large scales by continuum fluid mechanics, computer simulations of populations of air molecules could play an important role at these smallest scales.

The natural platforms for observing the turbulent structure of atmospheric variables are thus aircraft in the 'horizontal' and balloons in the 'vertical'. The turbulent structures of wind, temperature, and pressure themselves affect the motion of the platforms in air, preventing true motion confined to one or even two coordinates. Nevertheless, allowance can be made for such effects and the structure observed over five orders of magnitude in horizontal length scale and three orders of magnitude in vertical length scale, by in situ instruments along essentially $(1 + H), \approx 14/9$-dimensional aeroplane paths through the air (Lovejoy et al. 2004). Remote sounding, actively by lidars and passively by radiometers, has begun to be investigated. Sparling (2000) largely used satellite radiometric data to show that the probability distributions of atmospheric observations were not Gaussian. As data quality and resolution improve, such techniques could prove to be very fruitful, enabling in principle the application of multi-point correlation techniques (Shraiman and Siggia 2000) to the two-dimensional 'curtains' obtained under the aircraft and satellites carrying such instruments. Here, we confine our analyses to time series obtained by in situ instruments. We note that fractal, essentially Hurst exponent, time series analyses of surface observations of temperature have been performed by Koscielny-Bunde et al. (1998) and Syroka et al. (2001), and of ground-based total column ozone by Toumi et al. (2001). Another example is the establishment of power law correlations in column ozone abundance over Antarctica during the late winter–spring period of intense loss, the ozone hole (Varotsos, 2005). We do not consider such data here, because we are dealing with the free atmosphere rather than the surface, and because the frequency and resolution of such data is not generally adequate for examination of the phenomena in which we are interested here, although the approach could certainly be applied to higher frequency observations.

The instruments deployed aboard the aircraft described here are almost entirely of the in situ type, in which air flowing at about Mach 0.7 is decelerated when it enters an inlet and has some set of operations performed upon it to obtain the variable being measured. There are many aspects to this procedure, including the effects of physical configuration, control, electronic processing, and algorithmic analysis upon the time series which is ultimately recorded. Some but not all aspects of this can be tested—in many cases for example, Gaussian or Poisson noise is expected from the known instrument characteristics, or the time series can be analysed with and without 'spikes' in the time series arising from known interferences, such as plasma instabilities in the light source of the ozone instrument (Proffitt and McLaughlin 1983; Proffitt et al. 1989) or from cosmic ray impacts on the detector of the NO_y instrument (Fahey et al. 1989). Such effects have been successfully detected by the generalized scale invariance software (Tuck and Hovde 1999; Tuck et al. 2003b). While the internal consistency of the results between variables, flights, and missions to date is encouraging, it remains true that at some point both the internal design of the instruments and the response of the aircraft to the wind, pressure, and temperature fields will limit the interpretation of the meteorological data, particularly at the smaller scales.

2 Initial Survey of Observations

The observations are our starting point in this book, having been obtained from research aircraft in the last two decades. Justification for this approach can be found in Section 1.3 and by noting that there are no known analytical solutions to the Navier–Stokes equation, preventing the possibility of a priori prediction of the atmosphere's turbulent structure. We note the pioneering power spectral analysis of wind, temperature, and ozone from commercial Boeing 747 aircraft (Nastrom and Gage 1985) and the more recent data from Airbus 340 aircraft under the aegis of the MOZAIC programme (Marenco et al. 1998). Multifractal analysis was first applied to observations from an IL-12 aircraft in the tropics (Chigirinskaya et al. 1994) and has been applied to a large body of observations taken from ER-2, WB57F, DC-8, and G4 aircraft, with dropsondes from the last of these; Chapters 2, 4 and 5 are largely devoted to the results. Many of these data were obtained in the lower stratosphere from the ER-2 in the course of investigating ozone loss in both Arctic and Antarctic regions, where there exists a reasonably well-defined, durable circulation system offering clear dynamical, chemical, and radiative signatures. A more climate-driven imperative exists to investigate the tropical upper troposphere and lower stratosphere, largely pursued with the WB57F. The recent G4 and dropsonde data were acquired in the troposphere over the eastern Pacific Ocean, in the course of investigating northern hemisphere winter storms there.

2.1 An introduction to lower stratospheric research aircraft flights

The utility of balloons and then, 120 years later, from 1903, of powered aircraft for exploring atmospheric properties, were immediately obvious. The Second World War saw aircraft attaining stratospheric altitudes, revealing a very dry lower stratosphere with westerly winds in winter and easterlies in summer, with accumulation of high ozone abundances in polar regions (Brewer 1944; Dobson et al. 1945; Brewer et al. 1948; Brewer 1949; Murgatroyd and Clews 1949; Bannon et al. 1952). The fact that the shocked, hot air from nuclear weapons of greater than 1 MT yield stabilized above the tropopause, in the lower stratosphere, led to surveillance and monitoring programmes by high flying aircraft which established the basic mechanisms of dispersal, transport, and re-entry to the troposphere of dry, ozone-rich stratospheric air (Sawyer 1951; Murgatroyd et al. 1955; Murgatroyd 1957; Helliwell et al. 1957; Reed and Danielsen 1959; Feely and Spar 1960; Murgatroyd and Singleton 1961; Briggs and Roach 1963; Danielsen 1964; Murgatroyd 1965; Reed and German 1965; Danielsen 1968) and which were continued for two decades after the cessation of atmospheric testing by the USA, USSR, and UK in 1963 (Shapiro et al. 1980; Foot 1984), in part because of concerns about the effects of supersonic transport aircraft (Johnston 1971) and of halocarbons (Molina and Rowland 1974) on the integrity of the stratospheric ozone abundance. Over this time period, instruments for measuring temperature, pressure, and wind speed improved so that the data could be recorded at 1 Hz with adequate signal-to-noise ratios, enabling 'horizontal' spatial resolution of about 200 metres by an aircraft cruising at Mach 0.7. Such instruments were enhanced by the addition of instruments for ozone (Proffitt et al. 1989); water vapour and total water (Kelly et al. 1989, 1990, 1991, 1993); reactive nitrogen (Fahey et al. 1989) and condensation nuclei (Wilson et al. 1989) on to the NASA ER-2, a civilian version of the USAF's U2R reconnaissance aeroplane. So equipped and based in Darwin (12°S, 131°E) it investigated the tropical drying of air upon entry to the stratosphere (Danielsen 1993) and six months later, with chlorine monoxide (Brune et al. 1988) and nitrous oxide (Loewenstein et al. 1989) instruments added, successfully flew (Tuck et al. 1989) into the Antarctic ozone hole (Farman et al. 1985) from Punta Arenas (53°S, 71°W). This was followed by over a decade of missions with an ever more capable payload aimed at understanding the lower stratosphere and its relationship to the upper troposphere (Tuck et al. 1992; Anderson and Toon 1993; Wofsy et al. 1994; Tuck et al. 1997; Newman et al. 1999, 2002). The WB57F also undertook such missions from 1998, extending the ER-2 data into the upper tropical troposphere as well as the lower stratosphere (Tuck et al. 2003b; Richard et al. 2006). Because the stratosphere is largely stable in the vertical and has recognizable global scale horizontal flows, many of the flight legs on these missions were long great circle segments up to 7000 km in length, resolving over four orders of magnitude in horizontal scale at 1 Hz and nearly five at 5 and 10 Hz. The challenge is to produce more decades in horizontal length by speeding up the data collection frequency; however, this is made difficult by the conflicting requirements of signal-to-noise ratio and rapid instrument response. Finally, we shall see that the motion of the aircraft itself is affected by the structure of the atmospheric turbulence field (Vinnichenko et al. 1980; Lovejoy et al. 2004) and allowance has to be made for this effect.

Aircraft are obviously more suited to exploring the horizontal structure of the atmosphere than the vertical. Technological advances have made it possible to obtain more nearly vertical observations of horizontal wind speed, temperature, pressure, and relative humidity, particularly since the advent of the Global Positioning System, by dropping lightweight, disposable sondes from research aircraft (Hock and Franklin 1999; Aberson and Franklin 1999). When dropped from 13 km, a sonde takes over 800 seconds to splash down in the ocean while acquiring data at 2 Hz, so providing over 3 decades of vertical scaling. There are still many challenges in such observations, and improving the continuity by avoiding telemetry dropouts is important, as is the possible extension of the variables measured to ozone, aerosols, and possibly other radiatively important trace species. As with the aircraft, the sonde's motion is affected by the atmosphere's turbulent structure, which can be allowed for by modelling the Newtonian mechanics of the sonde as it falls under gravity balanced by aerodynamic drag.

Instruments on both aircraft and balloons have improved greatly for the staple measurements, such as pressure, temperature, winds, water, and ozone. The first three of these observations have been successfully performed at 5 Hz on aircraft and at 2 Hz on dropsondes, while the water and ozone have been so obtained at 1 Hz, with 2 Hz for relative humidity on the sondes. It should be noted that the electronic signals from the instruments are currently recorded at hundreds of Hz to 1 kHz, and are averaged to the lower frequencies given, which are the highest ones free of detectable Gaussian or Poisson noise. At some point, the residence time and turbulent structure of air in the instrumental detection volume imposes limits beyond which scaling analyses cannot be performed to get atmospheric rather than instrumental information; the limit can be both simulated by adding random noise to an atmospheric signal, and determined by examining it for changes in scaling behaviour, see for example Tuck et al. (2003b) and some figures included in Section 4.2. The limits above are what have been achieved to date; it is obviously more difficult for chemical species at low mixing ratios, where large, complex instruments are necessary to achieve signal-to-noise ratios above detectability.

2.2 A summary of the average scaling behaviour of in situ observations

The longest available 'horizontal' flight leg is a transatlantic flight of the NASA ER-2, which took place on 19890220 (yyyymmdd). The temperature trace at 1 Hz, equivalent to a spatial resolution of ~200 m, is shown in Figure 2.1. The corresponding wind speed trace is shown in Figure 2.2. The variability in both figures is almost entirely atmospheric; the effects of

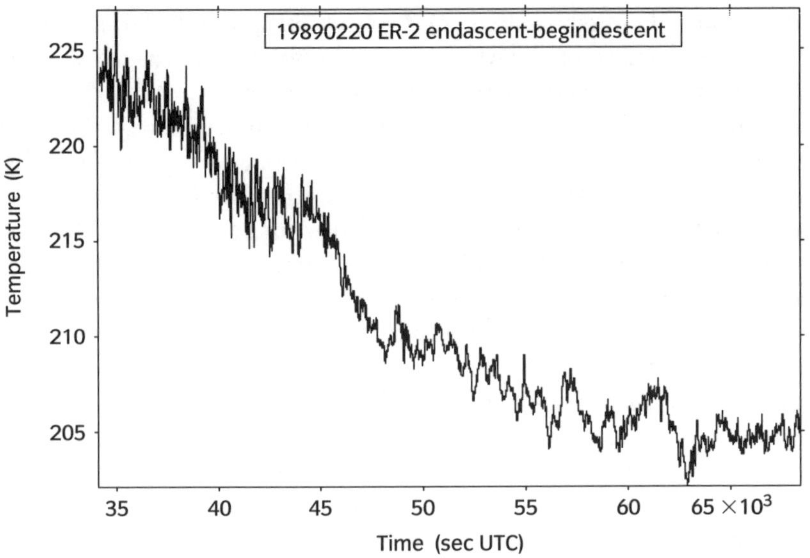

Figure 2.1 1-Hz ambient temperature data from the ER-2 flight of 19890220 [yyyymmdd], 20 February 1989, from the end of the ascent after take-off to the beginning of the descent before landing. The flight is the longest available horizontal segment, as the plane flew with a strong headwind in the outer vortex, between Stavanger (59°N, 6°E) and Wallops Island (38°N, 75°W) for 10 h 15 m. The variations are atmospheric, not instrumental; the record is one of a small fraction of the flights characterizing flight along rather than across the polar night jet stream.

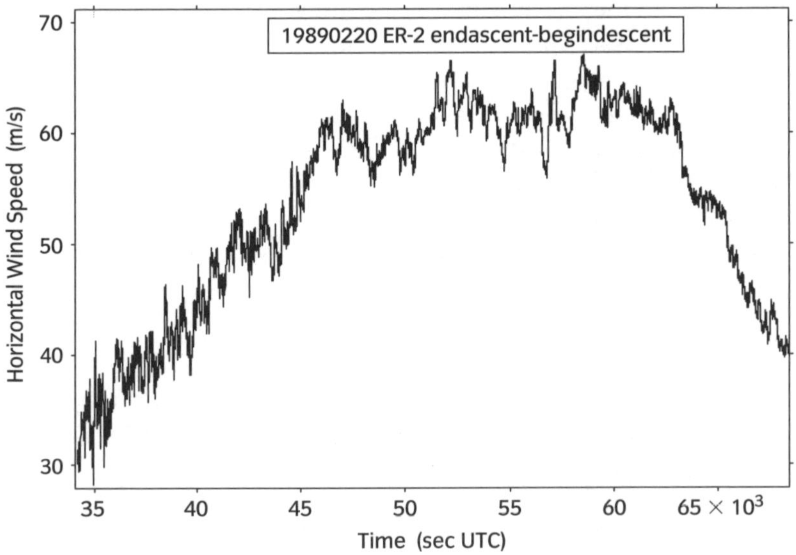

Figure 2.2 1-Hz horizontal wind speed data from the same ER-2 flight segment of 19890220, as in Figure 2.1. The fluctuations in wind speed are real and substantial in the along-jet direction.

instrumental (~Gaussian) noise are not evident until the data are analyzed at 10 Hz (Lovejoy et al. 2004). The wind shear vectors from the flight reveal that they essentially take all values between zero and 360°, but in an intermittent rather than in an ordered fashion; unidirectional flow is never encountered on any scale (Figure 2.3); the same is true for wind speed as it is for direction—speed shear is ubiquitous. These observations add a modern, data-based answer to Richardson's (1926) observation: 'Does the wind possess a velocity? This question, at first sight foolish, improves on acquaintance'. It does, but the averaging has to be carefully specified as a function of scale and time, which is what generalized scale invariance attempts to do in a compact manner, as discussed in Chapter 4.

The closest available approximation to truly vertical observation is from dropsondes; unlike data taken on ascent, there is no turbulent structure from the wake of the balloon. A sample of such data, acquired from a sonde dropped from the NOAA (National Oceanic and Atmospheric Administration) Gulfstream 4 over the eastern Pacific Ocean in 2004, is shown for temperature in Figure 2.4 and for wind speed in Figure 2.5. The contrast between the variability in the vertical profiles of temperature and of wind speed is immediately apparent. The temperature is smoother than the wind speed. There was no attempt to avoid dropping the sondes into convective cloud, although deep convection to the altitude of the aircraft was

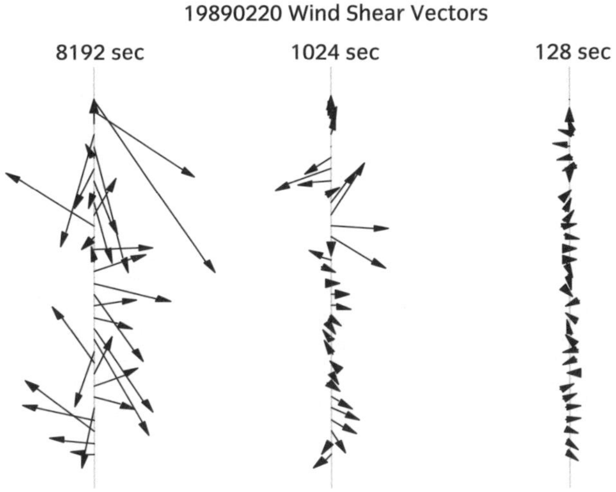

Figure 2.3 Horizontal wind shear vectors at three different time scales for the ER-2 flight of 19890220, for the same segment shown in Figure 2.1. The shear vectors show a wide range of speed differences among neighbouring intervals on all three scales. The differences in direction cover the whole 360°. The same behaviour is seen for all choices of scale, and implies that smooth, unidirectional flow at a uniform speed is non-existent on all scales. The scaling behaviour of many segments like that in Figures 2.1–2.3 is discussed in Chapter 4; the flight segment is used as an example of the horizontal observations.

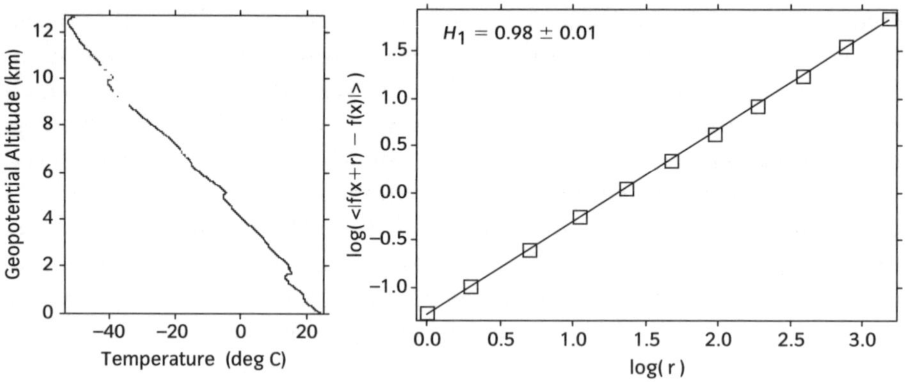

Figure 2.4 Temperature trace and variogram for the southernmost dropsonde during the Winter Storms 2004 mission, #27 on 04 March 2004 at (15° 15′ 11″ N, 165° 59′ 42″ W). The sonde was dropped from the NOAA Gulfstream 4SP research aircraft, which tracked its position as a function of time via the Global Positioning System (GPS) and received the telemetered data from the pressure, temperature, and relative humidity sensors. Note the smooth appearance of the temperature profile, accounting for the scaling exponent H_1 being close to but less than unity from the slope of the variogram, see Chapter 4.

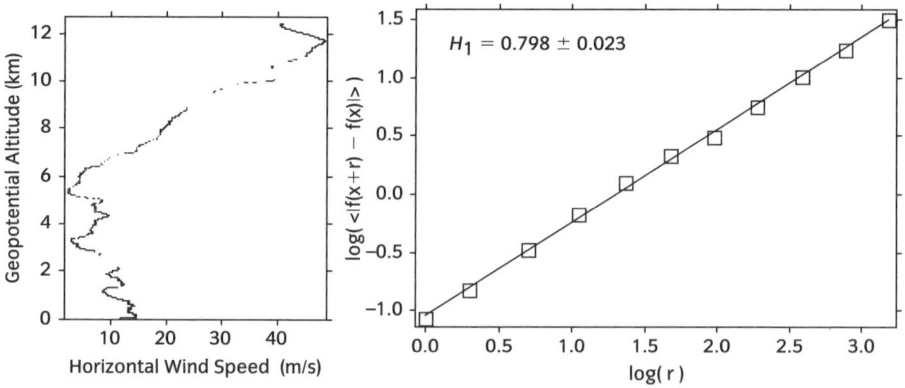

Figure 2.5 As Figure 2.4 but for horizontal wind speed, calculated from the time series of GPS positions. Note that the horizontal wind speed trace is rougher than that of temperature, with more variability on all vertical scales. As in the great majority of the dropsonde descents, there is a jet stream, shown by the wind speed maximum in the upper part of the profile. The H_1 scaling exponent is 0.8, significantly less than that for temperature in Figure 2.4, but more than the 0.6 predicted by Bolgiano-Obukhov theory, see Chapter 4.

not encountered often. Temperature clearly scales differently in the vertical than do wind speed and relative humidity, which scales like the wind speed. This result concerning the lapse rate probably reflects the direct influence of gravity, whereas the wind speed is affected by planetary rotation and horizontal pressure gradients and which in turn affects the distribution of water vapour. This will be discussed at greater length in Chapter 4.

Given the atmospheric variability shown above, we seek a mathematical framework to lend it coherence, by which is meant a compact description. Many trials with nearly 20 years worth of airborne data from four aircraft (ER-2, DC-8, WB57F, Gulfstream 4) have led us to use generalized scale invariance (Schertzer and Lovejoy 1985, 1987, 1991; Lovejoy et al. 2001, 2004; Tuck et al. 2002, 2003, 2004, 2005), in which observational time series are treated as statistical multifractals. We will examine detailed structure and individual flight segments later (Chapters 4 and 5); here we show composite variograms for a large volume of ER-2 'horizontal' aircraft data on temperature and wind speed in Figure 2.6. Equivalent results in the vertical for a large number of dropsondes appear in Figure 2.7. At this stage,

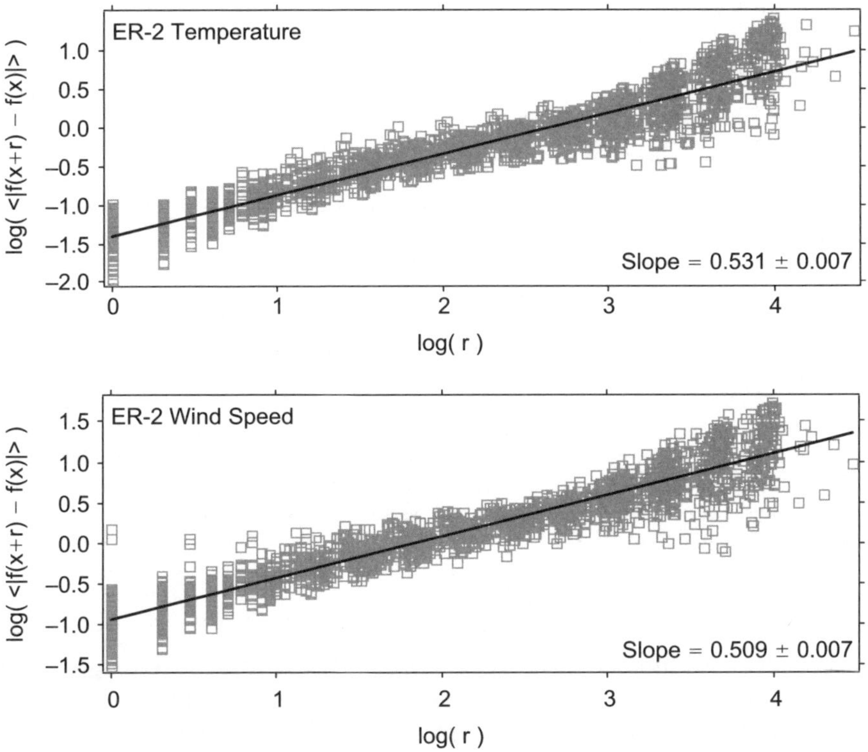

Figure 2.6 Composite variograms for temperature and wind speed measured along horizontal ER-2 flight legs. The data are all those collected on ER-2 missions, AAOE, AASE I, AASE II, SPADE, ASHOE/MAESA, STRAT, POLARIS, and SOLVE between 1987 and 2000 and apply to all legs under autopilot cruise of greater duration than 1800 seconds. The points obtained from each flight are plotted individually and the slope taken from all the points; there are many millions of 1 Hz observations contributing to the plot, from over 140 flight segments. The mean slopes yield values of H_1 close to the 5/9 predicted by generalized scale invariance, given that even in 'horizontal' flight the aircraft samples both horizontal and vertical structure; possible residual effects of the wind and temperature structure on the aircraft motion may account for the small differences (Lovejoy et al. 2004).

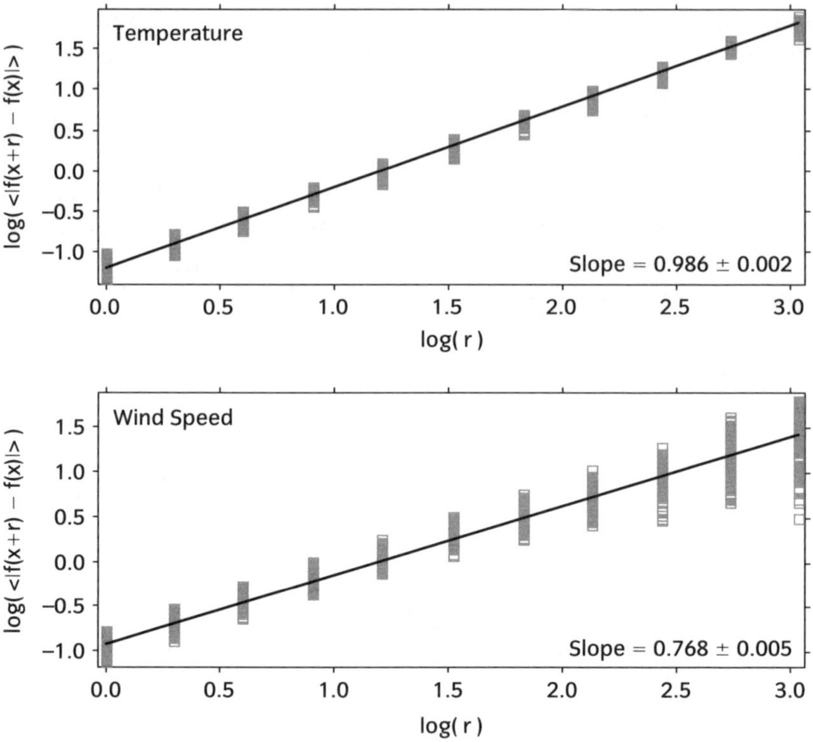

Figure 2.7 Composite variograms for temperature and wind speed measured by all drop sondes in the Winter Storms 2004 mission. These observations were taken over a large area of the eastern Pacific Ocean between 10°N and 60°N in late February to early March 2004, during 10 flights of the NOAA G4SP, and apply to the lowest 13 km of the atmosphere as observed by 261 sondes. The H_1 value of almost unity for temperature reflects the large influence of gravity through the hydrostatic equation in the vertical; evaluation by spectral analysis rather than the first order structure function yields a value of 1.25. Whichever method is used, temperature scales differently than wind speed or relative humidity in the vertical, which have H_1 values close to 0.75.

we will do no more than observe that the linearity of the log-log plots in these figures is impressive, with slopes yielding scaling exponents precise to better than 2%. Chapter 3 will discuss the underlying theory, but it can be concluded that statistical, multifractal scale invariance has been demonstrated for the kinds of observations shown in Figures 2.1–2.5, on scales from 40 m to 7000 km, or over 5 orders of magnitude down from an Earth radius.

If we plot the probability distribution functions for those same data, we do not find Gaussians; atmospheric data invariably have 'long' or 'heavy' tails. These are adjuncts of scale invariance, as we shall see in later chapters. There are many ramifications of this result, among them the occurrence of long-range correlations and influence upon the mean by a relatively small number of high amplitude events. A Gaussian is a sign that instrumental noise is dominating over atmospheric variability, as we can see in a few instances where the instrument noise, although small ($<7\%$ total) is larger than the atmospheric variability; water vapour in the lower stratosphere outside the tropics and the winter polar vortices has sufficiently low variability that this can be problematic. Another corollorary is that caution should be exercised in rejecting outliers in atmospheric data—there is no justification for forcing the observations into a Gaussian Procrustes' bed.

Reading

H. Kantz and T. Schreiber (1997), *Nonlinear Time Series Analysis,* Cambridge University Press, UK.

N. K. Vinnichenko, N. Z. Pinus, S. M. Shmeter and G. M. Shur (1980), *Turbulence in the Free Atmosphere,* Plenum Press, New York, 2nd edition. See Chapter 2 for a discussion of the aeroplane itself as a detector of turbulence.

3 Relevant Subjects

Atmospheric composition played an important part in the development of chemistry, following the work of Priestley, Lavoisier, and Dalton. Since air is a mixture of gases, many of them chemically reactive, see for example Finlayson-Pitts and Pitts (2000) and Graedel et al. (1986), which is subject to solar photons, absorbs and emits infrared photons, experiences temperatures ranging from -100 to $40°$ C, is exposed to the ocean, encompasses phase changes of water and sustains turbulent flow, it involves significant parts of physical chemistry. Pedagogically, the three-volume set by Berry, Rice, and Ross (2002a, b, c) covers the basic physicochemical material clearly and thoroughly, particularly Chapters 19, 20, 27, 28, 30, and 31. In addition to kinetic molecular theory, chemical kinetics, spectroscopy, and equilibrium statistical mechanics, there are other branches of physical science which are applicable to the atmosphere; in our context they include of course meteorology and turbulence theory. It ought to be recognized that the atmosphere has high complexity arising from a vast number of degrees of freedom, several anisotropies, and morphologically complicated boundaries extending over 15 orders of magnitude in scale from the molecular mean free path to the Earth's circumference; these factors and the concomitant non-linearities make the application of non-equilibrium statistical mechanics a daunting prospect, but nevertheless one which should be attempted, for the reason that the energy distributions and their transformations in the atmosphere need to be accurately described, particularly in the representation and prognosis of the climatic state. We will also show that vorticity is the fundamental variable, since vortices are generated from molecular populations subjected to an anisotropy, on very short space scales and fast time scales. In this Chapter we will give a skeletal survey connecting these basic subjects, with references to more comprehensive, individual sources.

3.1 Kinetic molecular theory

The simplest possible molecular model for a gas is a collection of spherical 'billiard balls'—the intermolecular potential consists of an infinite repulsive force on contact. This approach, pioneered by Waterston, Maxwell, and Boltzmann, is successful for air as a first approximation. The idea is that collisions are completely elastic, with no interaction between potential collidant molecules until physical contact occurs, whereupon an infinite repulsive potential operates. Compared to real nitrogen and oxygen molecules, this might seem to be too simple to work, and indeed it is when experiments are done that require a quantum mechanical description of the rotations and vibrations of these diatomic molecules. The two rotational degrees of freedom in N_2 and in O_2 are important in accounting for the difference between the specific heats of air at constant pressure and at constant volume. In real molecules, the existence of an attractive potential at longer ranges both increases the number of intermolecular encounters, and makes their definition less simple. However, the Maxwellian concept has proved to be a powerful means of formulating and understanding, at least to first order, the behaviour of large populations of molecules approximated as 'billiard balls', on a basis which is necessarily statistical.

The mean square molecular velocity along one Cartesian coordinate is

$$\overline{v_x^2} = \sqrt{\frac{m}{2\pi k_B T}} \int_{-\infty}^{\infty} v_x^2 e^{-mv_x^2/2k_B T} dv_x = \frac{k_B T}{m} \tag{3.1}$$

and in three dimensions the total kinetic energy of N classical particles is $3Nk_B T/m$ where m = molecular mass and k_B is Boltzmann's constant.

From here, the traditional approach to colligative behaviour was to calculate transport coefficients for an ideal gas slightly perturbed from equilibrium. We will not spend time on this, since the atmosphere is far from equilibrium, but give references: Hirschfelder, Curtiss and Bird (1964); Chapman and Cowling (1970). The original work was due to Chapman (1916), Enskog, Onsager, and Prigogine. See van Kampen (2002) for a current summary. Berry, Rice, and Ross (2002a, b, c) deal with the basic principles of how the Navier–Stokes equation can be derived for a molecular fluid and give useful examples; none of these books, with treatments aimed at those needing a description of laboratory scale flows, approach the range of scales and complexity present in the atmosphere, however. The subject will recur in Section 3.4, below. One result we take from Chapman and Cowling is the persistence of molecular velocity after collision, their pp. 93–96 and 327. It means that the assumption of no correlation in speed and position before and after collision with another molecule, which underlies Gaussian velocity distributions and Lorentzian line shapes in molecular spectra, is not valid. The expression is

$$\bar{w}_{12} = \frac{1}{2}m_1 + \frac{1}{2}\frac{m_1^2}{\sqrt{m_2}} \ln\left(\frac{\sqrt{m_2}+1}{\sqrt{m_1}}\right), \tag{3.2}$$

where $\bar{w}_{12}$ is the persistence ratio, the ratio of the mean velocity after collision to velocity before collision, m_1 is the mass of molecule 1 and m_2 is the mass of molecule 2. For equal masses, it has the value 0.406; heavy molecules take more collisions to lose their translational energy than light ones.

The central development was the invention of molecular dynamics simulation by numerical process on computers (Alder and Wainwright 1970), when it was shown how fluid mechanical behaviour emerges from a population of atomic scale elastic spheres. It was discovered that when an anisotropy (in the form of a pulse of fast molecules, or a flux) was applied to an equilibrated population of Maxwellian molecules, 'ring currents' evolved on a very short time scales (10^{-12} s) and on very small space scales (10^{-8} m). Figure 3.1 shows the original diagram from Alder and Wainwright's paper; the 'ring current' is hydrodynamic behaviour, what a meteorologist would call a vortex. A non-equilibrium statistical mechanical explanation was provided quickly by Dorfman and Cohen (1970); they showed that the molecular velocity autocorrelation function had a 'long tail', obeying a power law rather than an exponential decay at long times. Quantitatively, they found that the molecular velocity correlation function, C, was expressible as a function of the velocity $\boldsymbol{v}(t)$ at time t in systems of physical dimensionality d ($d = 2 \Rightarrow$ discs; $d = 3 \Rightarrow$ spheres) by

$$C(t) = d^{-1} \langle \boldsymbol{v}(t) \cdot \boldsymbol{v} \rangle \propto t^{-d/2}. \tag{3.3}$$

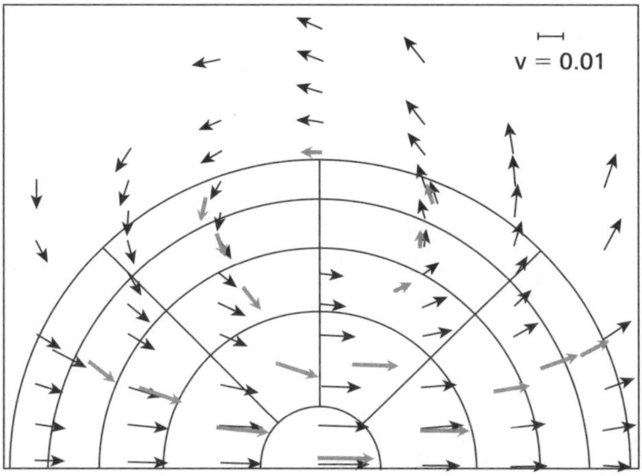

Figure 3.1 Alder and Wainwright's original molecular dynamics (MD) simulation of initially equilibrated Maxwellian atoms with an applied anisotropic flow. Thin black arrows represent simulation by the Navier–Stokes equation, grey arrows represent averages over the molecular velocity vectors after 9.9 collision times. Some later simulations show disagreement between MD and N-S calculations. The crucial point is that fluid mechanical behaviour has emerged from a randomly moving atomic population subjected to a flux.

Physically, the mechanism for the emergence of the vortices is a nonlinear interaction among the faster-moving molecules: they create high number densities ahead of themselves, leaving low number density behind. The number densities in these regions tend to equalize, creating the vortex flow. It is a point of central importance that the energies of the fast molecules and of the vortex feed into each other—they are mutually sustaining, via positive feedback. This is a new light on the concept of vorticity, suggesting that it can emerge under anisotropy, a flux, in the simplest possible representation of a population of molecules. It does so moreover on very short time scales, 10^{-12} seconds and on very short space scales, 10^{-8} metres; these scales are far smaller than the millimetric to centimetric scales at and below which true molecular diffusion has been traditionally considered to supplant fluid behaviour in the atmosphere. It has been suggested that a route from molecules to fluid mechanics could be obtained directly (Tuck et al. 2005) by taking the curl of the field of molecular velocities; in such an operation averaging procedures would be critical in proceeding from a granular to a continuum framework. Could turbulence be viewed as the emergence of larger scale, ordered motion from the more (but not completely) random motion of molecules? We will return to this in Section 7.1.

We will introduce a third important concept in this section, that of the fluctuation-dissipation theorem. 'Fluctuation' will be identified with the 'correlated' motions described above, in which the fast molecules in the PDF generated a vortex, which in turn sustained their excess translational energy. 'Dissipation' will be identified with the random behaviour of the molecules in the most probable region of the PDF (Probability Distribution Function) (it is the latter which permits an operational definition of a temperature in a non-equilibrium gas like the atmosphere, incidentally: see Sections 3.4, 5.2, and 7.1).

Historically, Einstein and Smoluchowski treated the Brownian motion of a particle immersed in a fluid, using dynamical and statistical methods respectively (1905, 1906). In 1908 Langevin linked the two treatments by handling the force exerted by the fluid molecules on the particle as the sum of its average value and the fluctuations about the mean. The former was handled dynamically, the latter statistically via probability theory. The Langevin equation:

$$m\frac{dv}{dt} = -\kappa v + \delta F(t), \tag{3.4}$$

where m is the particle mass, v its velocity, κ is kinematic viscosity, and $\delta F(t)$ is a fluctuating force, is the first fluctuation-dissipation theorem, expressing the fundamental relationship between the systematic and random motions in a dynamical system. It has many manifestations, both microscopic and macroscopic.

The conceptual models and associated mathematics generated in proceeding from kinetic molecular theory to fluid mechanics can become extremely complicated, involving many simplifying assumptions. A detailed account can be found in Chapman and Cowling (1970), and another specifically of the Boltzmann equation in Harris (1971); recent, more concise treatments may be found in Dorfman (1999), Succi (2001), and Zwanzig (2001). A full quantum mechanical treatment of the atmospheric population of molecules will remain a remote fantasy for the foreseeable future, for many reasons well known in non-equilibrium statistical mechanics and also because atmospheric molecular populations have insufficient time to achieve equilibrium distributions. This is so because the combined effects of solar flux, photochemistry, planetary rotation, and gravity induce turbulent vorticity structures, with non-equilibrated, long-tailed molecular speed PDFs, on short time and space scales. There will be positive feedback between the latter and the core of jet streams, large-scale structures which can have wind speeds that are a significant fraction of the average molecular speed.

If the density in phase space for a single molecule of mass m, position $\boldsymbol{r}$ and velocity $\boldsymbol{v}$ at time t is f, then Boltzmann's equation is

$$\frac{\partial f}{\partial t} + \boldsymbol{v} \cdot \nabla_r f + \frac{1}{m}\boldsymbol{F}(\boldsymbol{r}) \cdot \nabla_v f = \left(\frac{\partial f}{\partial t}\right)_{\text{collision}}, \tag{3.5}$$

where $\boldsymbol{F}(\boldsymbol{r})$ is an external force acting on the molecule. The left hand side represents the Liouville operator expressing Hamiltonian dynamics in a potential, and is time symmetric. The right hand side is the collision integral, and even for the simplest molecular model has squared terms in f; for a population of molecules it also contains probabilistic expressions. The equation has lost time symmetry, a concept examined via the H-equation in the above references; the squared terms mean that part of the equation does not reverse sign when t is replaced by $-t$. Equation (3.5) is one of many examples of fundamental equations describing the time evolution of samples of matter which look beguilingly simple in their derivation and in their compact expression, but which nevertheless are impossibly difficult to solve analytically. It is as true on the molecular scale as it is on the macroscopic, fluid scale for the atmosphere; the non-linearities and scale requirements can only be tackled by numerical methods. Fortunately, the continuing evolution of the memory and speed of electronic computers makes it much more possible to contemplate at least partial bridging of the gap between the bottom-up, microscopic approaches and the top down, macroscopic ones than was possible even a few years ago.

Reading

Berry, R. S., Rice, S. A., and Ross, J. (2002), *Physical Chemistry*, 2nd edition, Oxford University Press, Oxford.

Dorfman, J. R. (1999) *An Introduction to Chaos in Nonequilibrium Statistical Mechanics*, Cambridge University Press, Cambridge.
Harris, S. (1971) *An Introduction to the Theory of the Boltzmann Equation*, Holt, Rinehart and Winston, New York (re-issued by Dover Books, 2004).
Mazo, R. M. (2002) *Brownian Motion: Fluctuations, Dynamics and Applications*, Oxford University Press, Oxford.
Liboff, R. L. (2003) *Kinetic Theory: Classical, Quantum and Relativistic Descriptions*, 3rd edition, Springer, Berlin.
Zwanzig, R. (2001) *Non-Equilibrium Statistical Mechanics*, Oxford University Press, Oxford.

3.2 Turbulence

Historically, turbulence has had few connections to molecular approaches, Section 3.1, but a great many to fluid approaches, Section 3.3. In atmospheric work, it has been more applied in the boundary layer and in the very highest levels of the atmosphere, and in between has largely but not wholly been dealt with, one feels somewhat reluctantly, as part of the subgrid scale parameterization problem in numerical models for weather and climate. The reluctance is understandable; turbulence is a notoriously difficult subject. However, the importance of the transfer properties of smaller scale atmospheric turbulence to larger, 'meteorological', scales was evident, and clearly stated over half a century ago (Eady and Sawyer 1951).

Leonardo da Vinci studied turbulence, and actually used the word 'turbulenza' in a document dating to 1500, noting such phenomena as its generation, decay, and intermittency accompanied by drawings illustrating a range of scales in eddies, and even hinting at a separation into mean and eddy flow. Reynolds established from careful experimentation with flow along a pipe that there were two flow regimes, laminar and turbulent, and that the transition between them could be characterized by the dimensionless number Re where

$$Re = Lu/\kappa, \tag{3.6}$$

where L is a characteristic length scale of the flow, u is a characteristic velocity of the flow, and κ is the kinematic viscosity. If no care was taken in managing the flow, the onset of turbulence occurred when the laminar flow reached $Re \sim 2000$. However, if precautions were taken to eliminate perturbations near the pipe entrance, the transition occurred at $Re \sim 10^4$. Re for the atmosphere is $\sim 10^{11}$ in the boundary layer and $\sim 10^{12}$ in the free atmosphere. The wind field is turbulent by very large margins on this basis alone; we shall see that this is indeed the case on all scales, down to the smallest that can be measured with current technology.

In our meteorologically orientated development of the basic landscape, the next landmark was Richardson's (1926) experimental discovery that the mean square separation of two particles in atmospheric flow related separation distance x to time t_x by

$$t_x \propto x^{2/3}, \tag{3.7}$$

or that the rate of increase of the square of their separation distance varied as $x^{4/3}$, a power law dependence. This also inspired Richardson's well known adaptation of Swift's poem, and was the basis of Kolmogorov's 1941 formulation of decaying isotropic turbulence: see, in english, Kolmogorov (1991). The physical notion is that energy cascades down to smaller and smaller scales, being presumed to have been deposited on the largest possible scale. Independently of Kolmogorov, Onsager (1945, 1949), Heisenberg (1948) and von Weizsäcker (1948) deduced a similar result, the celebrated $k^{-5/3}$ law for the energy spectrum $E(k)$, where k is wave number and ε is the energy dissipation parameter:

$$E(k) \propto \varepsilon^{2/3} k^{-5/3}. \tag{3.8}$$

See Frisch (1995) for an historical account. The basic assumptions are that at high Reynolds numbers the statistical properties on scales smaller than the largest are solely and everywhere determined by the length scale, the energy dissipation rate, and the viscosity. As with many physical laws, this is an exact result, but for a very simple conceptual model. More realistic and therefore more complicated conceptual models engender a lack of mathematical exactness and the atmosphere is no exception. Gravity, planetary rotation, radiative heating and cooling, not to mention surface topography over both land and ocean, are all important. Seen from a basic physicochemical perspective, the fact that energy is deposited in the atmosphere by molecules absorbing photons suggests that conservative energy cascades are first of all unlikely, and secondly would have to be upscale rather than downscale. The scale invariant, turbulent fluctuations in the abundances of the absorbing molecules such as oxygen, ozone, water, carbon dioxide, methane, and nitrous oxide ensure that energy will be input to the atmosphere on all scales, whether it is solar or terrestrial radiation. We shall see that there is observational evidence for this view in later chapters.

Landau and Lifshitz (2003) point out a longstanding (1944) objection to the universality of the $k^{-5/3}$ law, namely that what is now called intermittency invalidates, through similarity assumptions built into the averaging, the constancy of x in Equation (3.7): x will be scale dependent. There are relevant discussions in section 6.1.3.4 of Davidson (2004) and Chapter 8 of Frisch (1995). The accommodation of the intermittency observed in the atmosphere necessitates more than one scaling exponent (the power spectral exponent β, which is $-5/3$ in the Kolmogorov theory, does not suffice) and treatment on a statistical and multifractal basis, as we shall see in Chapter 4. The scale dependence of x means that scale invariance has not been destroyed, but has become anisotropic, multiscaling and requiring three exponents to describe the scaling.

The intermittency observed in shear flow by Batchelor and Townsend (1949), was what spurred Mandelbrot (1974, 1998) to originate the theory of multifractals, a term coined by Parisi and Frisch (1985). The formulation for and application to atmospheric observations was by Schertzer and Lovejoy (1985, 1987, 1991). There is a large literature on multifractality generally, spread over many scientific disciplines and using varied terminology and choice of symbols. The atmospheric literature is of manageable dimensions however; we adhere to the Schertzer and Lovejoy formulation. A detailed account is in Chapter 4 of *Nonlinear Variability in Geophysics*, Schertzer and Lovejoy, eds. (1991); a more accessible and more meteorological account is in Schertzer and Lovejoy (1987). The basis is a statistical treatment of multiplicative random processes, leading, it is argued, to a complete description by three scaling exponents. These are H_1, the conservation exponent; C_1, the intermittency; and α, the Lévy exponent. H_1 is a measure of the degree of correlation: in the limit of zero intermittency, $H_1 \rightarrow 1$ corresponds to perfect neighbour-to-neighbour correlation, $H_1 \rightarrow 0$ corresponds to complete anti-correlation, that is Gaussian noise. The intermittency C_1 is the co-dimension of the mean of the field, that is a measure of its sparseness. The Lévy exponent α measures the power law fall-off of the tail of the probability distribution. Their ranges are $0 < H_1 < 1$, $0 < C_1 < 1, 0 \leq \alpha \leq 2$. Experience with high resolution in situ observations suggests that $H_1 \sim 5/9$ for passive scalars and wind speed, while total water, ozone, reactive nitrogen, and chlorine monoxide can vary through the operation of sources and sinks; indeed this use of the H exponent is a numerical model-independent way of deducing the existence of non-conservative processes. Where the data are good enough, the intermittency for wind and temperature tends to be in the range $0 < C_1 \leq 0.10$, which although at the low end of the possible range are nevertheless significant values. Accurate determination of α demands large volumes of precise, gap-free data, and really needs better instruments, in the sense of faster and gap-free performance yielding higher volumes of data, than we have currently. $\alpha \approx 2$ corresponds to Gaussian noise; what determinations there are (for winds, temperature, and ozone) suggest $1 < \alpha < 2$. The PDFs associated with $H_1 \approx 0.55$, $C_1 \approx 0.05$ and $\alpha \approx 1.6$ are obviously asymmetric, with 'long' tails. Finally, for a prescient view of large-scale atmospheric turbulence, see Eady (1950); for a recent view from the bottom up, see Tuck et al. (2005). It is appropriate to finish with the further words of Eady (1951).

> 'I congratulate Dr. Batchelor on his scholarly presentation of the similarity theory of turbulence initiated by Kolmogoroff. The argument which derives the consequences of statistical "de-coupling" between the primary

> turbulence-producing processes and the secondary small-scale features of the turbulence appears to be sound but does it get us very far? In meteorology and climatology we are concerned principally with the transfer properties of the turbulence, determined mainly by the large-scale primary processes to which the similarity theory does not pretend to apply. It is the great virtue of similarity theories that no knowledge of the mechanism is involved and we do not have to assume anything about the nature of "eddies"; anything which has "size" (such as a Fourier component) will do in our description of the motion. But this emptiness of content is also their weakness and they give us very limited insight. It is true that a similarity theory that could be applied to the *primary* turbulence-producing processes would be of great value but there is no reason to expect that anything simple can be found; when several non-dimensional parameters can be formed, similarity theory, by itself, cannot do much.
>
> Similarity theories are attractive to those who follow Sir Geoffrey Taylor in rejecting crude hypotheses regarding "eddies", mixing lengths, *etc.* But those who try to determine the properties of turbulence without such (admittedly unsatisfactory) concepts must show that they have sufficient material (in the shape of equations) to determine the answers. If this is not the case it will be necessary to develop some new principle in addition to the equations of motion and the nature of this principle may be brought to light in a study of the mechanism of the primary turbulence-producing process *i.e.* by trying to refine or modify what we mean by an "eddy" rather than by completely rejecting the concept.'

A wider context for the importance of understanding the mechanisms of turbulence can be found in Eady and Sawyer (1951).

Will the flux-driven emergence of hydrodynamics on molecular scales discovered by Alder and Wainwright (1970), followed by computer simulation of upscale propagation of vorticity structures to the jet stream scale via overpopulations of high speed molecules, provide the turbulence-producing process and eliminate the emptiness of content described in Eady's comment? Defining an eddy amounts to defining turbulence and is probably best avoided. Note that an 'eddy' is a fluctuation, in the sense of being a departure from the mean, per Equation (3.4), with the mean representing 'organization' or 'order' in the Langevin approach. This contrasts with the view arising from a molecular dynamics—non equilibrium statistical mechanics approach, in which the overpopulation of fast molecules generates the fluctuations, vorticity structures, or ring currents, representing order and the dissipation is represented by the random motions of more average molecules with velocities near the most probable.

Reading

The books by Davidson (2004) and Frisch (1995) are highly recommended, particularly the former. The second edition of the book by Vinnichenko et al. (1980) gives an account of the classical approach to atmospheric turbulence.

3.3 Fluid mechanics

Prompted by the spontaneous production of vortices on very short time scales and very small space scales in molecular dynamics computations described in Section 3.1, we will view vorticity as the fundamental variable. See Davidson (2004), sections 2.3–2.4 for a macroscopic discussion, which is even brave enough to offer a definition of turbulence! The use of 'chaotic' in it, however, may need modifying in the light of the operation of fluctuation-dissipation theorems. The competition between 'order' (fluctuations, arising from correlations) and 'disorder' (dissipation arising from random motions) cannot be avoided in air on any scale. It fundamentally limits Lagrangian approaches and the use of chaos theory, which depend on conservative movement of a 'particle' in physical space and phase space respectively. Dissipation ultimately occurs conceptually when energy is redistributed among air molecules according to the Maxwell-Boltzmann distribution of molecular speeds, Equation (3.1), a state corresponding to thermodynamic equilibrium. While this may be a workable approximation in some circumstances, and indeed the equilibration of kinetic energy among the majority of molecules close to the most probable and mean speeds in one or two collisions is what makes temperature definition and measurement operationally possible, we have rehearsed arguments in Section 3.1 that thermalization will not be complete in the atmosphere, a phenomenon leading to the rapid, very small scale generation of vorticity via the overpopulation, relative to a Maxwellian distribution, of faster-moving molecules.

The vorticity form of the Navier–Stokes equation in three dimensions (note that observationally $H_1(s) \neq 1$ or zero, where s is horizontal wind speed, so we cannot expect to view atmospheric vorticity in two dimensions and remain quantitative, since the dimensionality of atmospheric flow is $2 + H_1(s)$, see for example Schertzer and Lovejoy (1985, 1991) and Tuck et al. (2004)) is

$$\frac{D\omega}{Dt} = (\boldsymbol{\omega} \cdot \nabla)\mathbf{u} + \kappa \nabla^2 \boldsymbol{\omega}. \tag{3.9}$$

The first term on the right says that vorticity, $\boldsymbol{\omega}$, advects itself: nonlinearity is inherent. This term alone is responsible for much of the complexity and difficulty associated with understanding, describing and computing atmospheric flow under the continuum assumption.

$\boldsymbol{\omega}$ is defined by

$$\boldsymbol{\omega} = \nabla \times \mathbf{u} \tag{3.10}$$

and by using the autocorrelation function for vorticity

$$\langle \boldsymbol{\omega}(t) \cdot \boldsymbol{\omega}'(t) \rangle = -\nabla \langle \boldsymbol{u}(t) \cdot \boldsymbol{u}'(t) \rangle = C(t) \tag{3.11}$$

where t is time, we can define enstrophy:

$$\mathscr{E} = \frac{1}{2}|\boldsymbol{\omega}|^2 = 2C(t) \tag{3.12}$$

and enstrophy is governed by

$$\frac{D}{Dt}\left(\frac{|\omega|^2}{2}\right) = \omega_i \omega_j S_{ij} - \kappa|\nabla \times \boldsymbol{\omega}|^2 + \nabla \cdot [\kappa \boldsymbol{\omega} \times (\nabla \times \boldsymbol{\omega})] \tag{3.13}$$

This expresses the generation of vorticity by stretching, or its destruction by compression, via the first term, balanced by viscous dissipation in the second term. The third term is the divergence, often assumed to be locally zero; this cannot be strictly true, for example, if ozone photodissociation is leading to the generation of vorticity. Here S_{ij} is the straining rate on a fluid element.

Statistical mechanics (Sections 3.4 and 7.1) offers several definitions of entropy. One uses vorticity, see for example Bell and Marcus (1992). If P_l is the probability density in the lth bin (grid box, cell, pixel ...) of size δ_ω, and there are L bins all of size δ_ω, with
$\omega_l = l\delta\omega$, and $l = 1, \ldots, L$,

$$P_l = \sum_{ijk} \mu(\omega_{ijk}, l) \tag{3.14}$$

where $\mu(\omega_{ijk}, l) = N^{-3}$ if $|\omega_l - \omega_{ijk}| \leq \delta_\omega/2$ and zero otherwise.

An entropy of the vorticity field is then given by

$$S_w = -\sum_l P_l \ln P_l \tag{3.15}$$

(not to be confused with S_{ij} above).

We thus can associate entropy with vorticity, an alternative to potential temperature. Because the macroscopic definition of temperature is

$$\frac{dS}{dE} = \frac{1}{k_B T} \tag{3.16}$$

(see Section 5.2), vorticity can be related to the thermodynamic state of the flow (Truesdell 1952, 1954) as originally done by Beltrami (1871).

Vorticity can be normalized over an air column by dividing by the depth of the column, to yield a conservative quantity, potential vorticity (Rossby 1940; Ertel 1942; Hoskins et al. 1985). The full development in meteorologically familiar notation is in the last of these three references, which expounds 'PV thinking'.

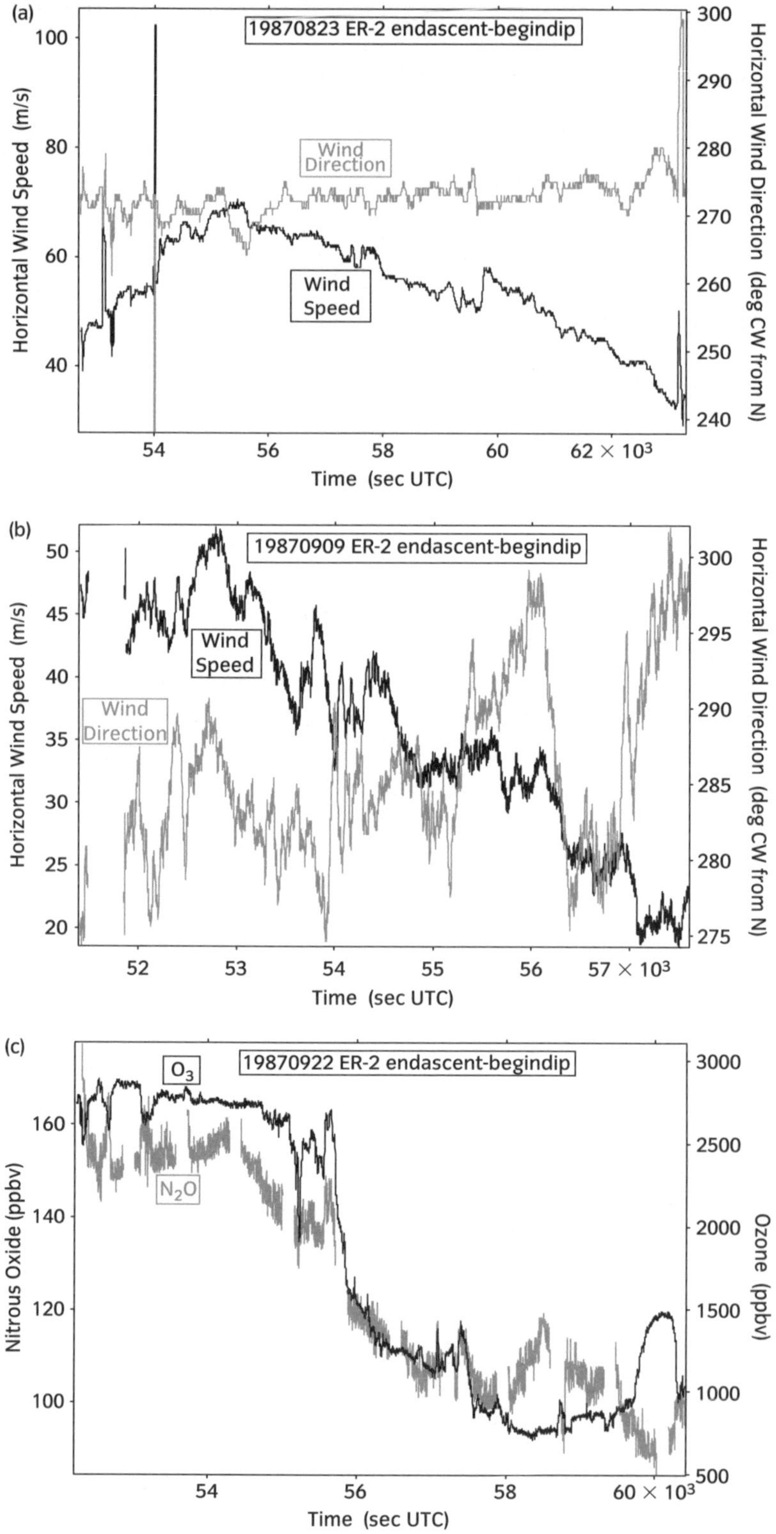

Figure 3.2 *Continued.*

For isentropic flow, using potential temperature θ—which physically is the temperature an air parcel would attain by compression if brought adiabatically to the surface—as the vertical coordinate, the absolute vorticity $(\zeta + f)_\theta$ is the sum of the relative vorticity ζ and the planetary vorticity f, so

$$\frac{d}{dt}(\zeta + f)_\theta = -(\zeta + f)_\theta \nabla_\theta \cdot \boldsymbol{v} \tag{3.17}$$

By employing the hydrostatic assumption, pressure p and θ can be related and the equation re-written as

$$\frac{d}{dt}\left\{(\zeta_\theta + f)\frac{\partial\theta}{\partial p}\right\} = 0 \tag{3.18}$$

and

$$Q_\theta = (\zeta_\theta + f)\frac{\partial\theta}{\partial p} \tag{3.19}$$

expresses the potential vorticity, Q_θ. It is the product of the inertial and convective stability terms; if either becomes negative instability ensues, if both terms retain the same sign, flow remains stable. Potential vorticity allows the use of vortex stretching in examining the Lagrangian movement of air. We will see later that all the basic measured quantities from which potential vorticity is computed—wind speed, temperature, potential temperature—are all scale invariant because all possess turbulent structure. There are thus limits to the time scales upon which potential vorticity can be used as a Lagrangian tracer; it is more useful on synoptic meteorological scales of a few days than on longer ones or for describing climate. Like everything else involved in atmospheric dynamics, it is subject to the fluctuation-dissipation

---

Figure 3.2 (a) Horizontal wind direction and wind speed traces from the ER-2 flight of 19870823 from Punta Arenas (53°S, 71°W) to a point just south west of Alexander Island (72°S, 80°W), end ascent to begin dip. (b) As for (a) but for the flight of 19870909 from Punta Arenas to (68°S, 68°W). (c) Ozone and nitrous oxide traces from the ER-2 flight of 19870922 from Punta Arenas to (72°S, 71°W), end ascent to begin dip; there is a random instrumental noise component in the nitrous oxide trace. Ozone and nitrous oxide are normally negatively correlated in the lower stratosphere on large scales because ozone has a photochemical source above, while nitrous oxide has a photochemical sink there. The special conditions in the ozone hole give ozone a local sink and a positive correlation with nitrous oxide, which has yet to fully develop locally between 58 000 and 60 000 seconds UTC. Note the intermittencies and steep gradients, characteristics associated with fractality and scale invariance. The associated PDFs have fat tails and are non-Gaussian. The ozone does not scale like a passive scalar (tracer), presumably because of the chemical sink in the vortex, see Sections 4.3 and 6.2. There are 86 400 seconds in a day, with zero being midnight GMT; local times were 18 000 seconds behind this, thus for example 54 000 s UTC is 15:00 hrs UTC and 10:00 hours local.

theorem; reference to the scale invariant wind and shear vectors in Figure 2.3 provides direct observational support for this statement.

We note that we now have a framework connecting molecular dynamics → vorticity → enstrophy → entropy → temperature. Temperature can also be defined in molecular terms, as we have seen in Section 3.1 and will revisit in Section 5.2. Atmosphere-specific considerations produce potential temperature ($\propto$ ln[entropy]), potential vorticity and potential enstrophy, and of course entail anisotropies in the form of gravitation, planetary rotation, the solar beam, and the surface. The third law of thermodynamics is finessed by normalizing pressure to 1000 hPa in the equation for potential temperature. Nevertheless, it is clear that vorticity is fundamentally related to, indeed emerges from, molecular behaviour. We further note that there is observational evidence for (a) atmospheric wind speeds that are a significant fraction of the most probable molecular velocity (130 ms^{-1} in the sub tropical jet stream vs. 390 ms^{-1} at 200K) and (b) extremely sharp gradients, see Figure 3.2, where a conserved tracer and wind speed show such a phenomenon in the Antarctic polar night jet stream. Conditions (a) and (b) violate the assumptions underlying the derivation of the Navier–Stokes equation, even in its compressible form.

The observation that jet stream core speeds can be a significant fraction of the most probable speed of air molecules leads naturally to numerical simulation of a high Reynolds number gas. This is a vast field, of which weather forecasting and climate simulation are but part, albeit a significant one. Here, we focus on what we have covered so far implies for parametrization at unresolvable scales in global scale models. Since we have observed $H_1(s) \neq 1$, it would be seem to be true that while macroscopic hydrodynamic stability analysis can predict, for example, that a westerly baroclinic current will be unstable, it will be less successful at predicting exactly where, when, and in particular, the detailed characteristics of the ensuing turbulence. However, the fact that molecules beget vorticity, and that we have observed scale invariance, described by a small set of scaling exponents, suggests a possible route forward. It will necessarily be stochastic, but statistical descriptions are at hand. The necessity of doing parametrization from the bottom up has been suggested by Palmer (2001); currently, it is a large uncertainty in climate models, where an accurate description of the energy state of the atmosphere is essential. Since a lot of the energy enters and leaves the air as photons absorbed and emitted by molecules, by definition the smallest scales, our arguments seem to be highly relevant, touching even the question of what atmospheric temperature really is, see Section 5.2. Even if one considers the energy transferred to the atmosphere from the Earth's surface, whether via radiative emission in the infrared followed by atmospheric absorption, thermally by conduction or by momentum transfer, it is still molecules moving, that is to say the smallest scales are operative.

3.4 Non-equilibrium statistical mechanics

Statistical mechanics has provided a beautiful, accurate, and highly successful framework for the description of matter at equilibrium. The treatment of matter far from equilibrium, such as the Earth's atmosphere, has been more problematic. The need to deduce and formulate turbulent flow starting with the Maxwell–Boltzmann–Gibbs probability distributions of molecular energy has been appreciated by relatively few scientists. Exceptions however included Grad (1958, 1983) and von Neumann (1963). Very recently, a whole book has been devoted to an attempt, via the definition of a 'turbulent Gibbs distribution' (Chen 2003). On the macroscopic level, the way has been pioneered by a meteorologist (Paltridge 1975, 2001) and recently put on a more secure quantitative basis by Dewar (2003, 2005a, 2005b) via the principle of maximum entropy production. The incoming beam of solar radiation from a black body at ~5800 K is a relatively low entropy state ($\delta S \approx \delta E/T$ is small) while the outgoing infrared radiation at ~255 K ($\delta S \approx \delta E/T$ is large) over 4π solid angle is a relatively high entropy state. The 'entropy dump' to space is what drives organized circulation in the atmosphere; there is a continual interplay between 'fluctuation' (correlated motion and solar photon absorption by molecules) and 'dissipation' (uncorrelated motion and infrared emission by molecules).

The planetary energy balance can be expressed by the well-known relation

$$(1 - a)I_o = \sigma T_R^4, \tag{3.20}$$

where a is the planetary albedo, I_o is the net incident flux of solar radiation, σ is the Stefan-Boltzmann constant, and T_R is the effective net radiative temperature of the planet. We can express the total entropy production using this relation as

$$\Delta S = (1 - a)I_o(1/T_R - 1/T_{SOLAR}) \tag{3.21}$$

which amounts in a long term global average to about 900 mW m^{-2} K^{-1} (Kleidon and Lorenz 2005). There is of course a great deal of fine structure and detailed physics underlying this relation! It is worth noting that the maximization of entropy production entails maximization of dissipation (Paltridge 2001, 2005), a result which would have pleased Eady (1950), relying as it does on turbulent transfer of heat. Paltridge (2005) also points out that the minimization of entropy exchange is equivalent to the maximization of entropy production. The Shannon information entropy view of this statement is that advection minimizes the number of coordinates needed to describe the air motion, whereas dissipation maximizes them. Anisotropies, often in the shape of fluxes, prevent complete dissipation to an isotropic state. The flux of heat (molecular velocity) is central.

The maximum entropy formalism depends upon Jaynes' information theory of non-equilibrium statistical mechanics, and as formulated by Dewar (2003, 2005a, 2005b) depends upon treating the phase space paths between two states so as to maximize the path information entropy $S_\Gamma = -\sum_\Gamma p_\Gamma \log p_\Gamma$ where p_Γ is the distribution of paths with non-zero probability. It is equivalent to what Jaynes called the caliber, and formally is the same as Gibbs's expression for equilibrium states. In a molecular context, Evans and Searles (2002) has shown that the decay of microscopic fluctuations deviating from the 2nd law of thermodynamics is exponential, a result which has passed experimental test (Wang et al. 2002). It should be noted that Jaynes's approach is still controversial, see for example Balescu (1997) Chapter 16.2, and Dougherty (1994).

As discussed by Kleidon and Lorenz (2005), meteorological objections to the applicability of maximum entropy production to the Earth's atmosphere have included the absence of the rate of planetary rotation rate from the expression for the heat transport, and the idea that the system 'wishes' to find the state of maximum entropy production. Paltridge (2005) refers to 'magical means' by which a non-linear system attains its state of maximum entropy production. We can make suggestions about the resolution of these difficulties as follows. We will see in Chapter 4 that the scaling exponent for wind speed, $H_1(s)$, is correlated with measures of jet stream strength in both the horizontal and the vertical. Since the scaling exponent is a reflection of the turbulent structure, and jet stream speed directly includes the Coriolis force in its expression in dynamical meteorology via the thermal wind equation, we conclude that the turbulent heat transport must depend upon the planetary rotation rate. The molecular view of non-equilibrium statistical mechanics can illuminate how the system finds the state of maximum entropy production: it is the same argument as is used in equilibrium statistical thermodynamics, that a large population of molecules has so many collisions that the motion of the molecules will sample enough of the accessible states that the most probable state will always be the result in a macroscopic system. The emergence of vortices on very short time and space scales in a population of Maxwellian molecules subjected to an anisotropy (Alder and Wainwright 1970) provides the mechanism in a non-equilibrium system—it is the basic causal process of the faster moving molecules interacting with each other to produce 'ring currents' and a non-thermal distribution of molecular speeds. The mutually sustaining interaction between overpopulations of fast molecules, relative to a Maxwellian, and vorticity structures noted previously in Section 3.1 constitute a mechanism by which the molecular behaviour could be propagated up to successively larger scales. The fact that core jet stream wind speeds can be a significant fraction of the most probable molecular speeds suggests that positive feedbacks between the overpopulated fast tail of the molecular PDF and the fast macroscopic flow are possible. Recall that we have suggested that the fluid mechanical vorticity structures represent an ordered state relative to thermalized molecules, and that the entropy production associated with the latter is what makes the existence of the former possible. In a flux-driven system, which is always true of the air, complete order would be represented by an organized flow, with all the molecules sharing a common direction and speed. This is not possible because of the fundamental dynamics of molecular collisions. Complete dissipation on the other hand would be represented by a universal, completely thermalized Maxwellian speed distribution. In this hypothetical state, nothing could happen, a kind of 'entropy death' for the planet; this is not possible in the atmosphere, which has been maintained far from equilibrium by the combined anisotropies of the solar flux, gravity, planetary rotation, and surface topography. There must be a maximum in between, corresponding to the continual observed competition between advection and diffusion. Paltridge (2001) showed that maximum dissipation corresponded to maximum entropy production. We will return to this subject in Chapter 7, to provide perspective after surveying observational analyses for scale invariance.

3.5 Summary

We have seen that large collections of time series of accurate, precise atmospheric observations which are otherwise unruly (in the sense of showing high amplitude, incoherent variations) can be made coherent by statistical multifractal analysis. The three exponents, H_1, C_1, and α, used in the analyses are associated with non-Gaussian PDFs, with long tails and with power law behaviour. The values of these exponents show that the turbulent structure attending atmospheric motion is neither 2D nor 3D, as we shall see in Chapters 4 and 5. From the molecular dynamics literature, we have seen vortices develop on very small spatial scales (10^{-8}m) and short time scales (10^{-12}s) in a gas of Maxwellian molecules to which an anisotropy had been applied. Vorticity is thus closely related to temperature, the average of molecular square velocities. From vorticity can be defined enstrophy, half the mean square vorticity; in turn, an entropy can be defined from enstrophy. Very recently, entropy has been related to scale invariance (Tsallis et al. 2005). Scale invariance could also arise from the properties of Chen's (2003) 'turbulent Gibbs distributions', see his p. 341. A connection between maximization of entropy production and the existence of scale invariance in turbulent flow would be of great conceptual importance. The disciplines of fluid mechanics and non-equilibrium statistical mechanics have yet to be merged in a rigorous framework capable of this linkage, however. Nevertheless, glimpses can be had of useful physical insight, and the use of a scaling exponent in a cloud microphysical model to impose realistic temperature fluctuations has been shown to cause substantially different ice cloud formation (Murphy 2003). Could it work for sub grid scale parametrization in climate models? The existence of scale invariance suggests the possibility of a computationally economical representation.

4 Generalized Scale Invariance

Probability distributions plotted to date from large volumes of high quality atmospheric observations invariably have 'long tails': relatively rare but intense events make significant contributions to the mean. Atmospheric fields are intermittent. Gaussian distributions, which are assumed to accompany second moment statistics and power spectra, are not seen. An inherently stochastic approach, that of statistical multifractals, was developed as generalized scale invariance by Schertzer and Lovejoy (1985, 1987, 1991); it incorporates intermittency and anisotropy in a way Kolmogorov theory does not. Generalized scale invariance demands in application to the atmosphere large volumes of high quality data, obtained in simple and representative coordinate systems in a way that is as extensive as possible in both space and time. In theory, these could be obtained for the whole globe by satellites from orbit, but in practice their high velocities and low spatial resolution have to date restricted them to an insufficient range of scales, particularly if averaging over scale height-like depths in the vertical is to be avoided; analysis has been successfully performed on cloud images, however (Lovejoy et al. 2001). Some suitable data were obtained as an accidental by-product of the systematic exploration of the rapid (1–4% per day) ozone loss in the Antarctic and Arctic lower stratospheric vortices during winter and spring by the high-flying ER-2 research aircraft in the late 1980s through to 2000. Data initially at 1 Hz and later at 5 Hz allowed horizontal resolution of wind speed, temperature, and pressure at approximately 200 m and later at 40 m, with ozone available at 1 Hz, over the long, direct flight tracks necessitated by the distances involved between the airfield and the vortex. Some later flights also had data from other chemical instruments, such as nitrous oxide, N_2O, reactive nitrogen, NO_y, and chlorine monoxide, ClO, which could sustain at least an analysis for H_1, the most robust of the three scaling exponents. Better than four decades of horizontal scale were available for 1 Hz and 5 Hz data. Since then, a lesser volume of adequate data has been obtained away from the polar regions by the WB57F. While there are other atmospheric research platforms actively flying in the troposphere, such as DC-8, WP3D, and G4 aeroplanes, they generally have pursued objectives which limit the length and duration of great circle flight segments. The NOAA G4 however, during dropsonde missions over the eastern Pacific, has not only flown such simple flight segments, it has dropped Global Positioning System (GPS) sondes which enable the scaling of horizontal wind speed, temperature and humidity to be examined in the vertical between 13 km and the surface over more than three decades of scale. We will concern ourselves with the application of the generalized scale invariance approach to the ER-2, WB57F and G4 data where appropriate and including the drop sonde observations.

4.1 Mathematical framework of generalized scale invariance

We start with PDFs of high altitude aircraft temperature data from the Arctic lower stratosphere and from the tropical tropopause regions. Figure 4.1 displays three such distributions; each has a highly asymmetric shape, with the winter vortex and the tropical tropopause region both having cold most probable values with a long warm tail, relative to the annual mean for the region, while the reverse is true for the Arctic summer anticyclone. The NCEP (National Centers for Environmental Prediction) analysis of the tropical tropopause temperature during boreal winter displays a similar structure to the observations, showing that for this region at least that asymmetric PDFs are global. The aircraft data apply of course to the geographical regions accessed by the flight tracks, but the sampling was frequent enough at intervals sufficiently far apart in time that the circumpolar airflow should counter the bias arising from the localized nature of the aircraft sampling. In a general sense of the morphology of the wind, potential vorticity, and trace species fields, this can be seen from meteorological analyses and satellite data.

A detailed exegesis of the mathematical theory underlying generalized scale invariance is available in the work of Schertzer and of Lovejoy referenced in both Chapter 2 and in this chapter. Here we will state a few basic relations and specify the procedure used to calculate the three scaling exponents H_1, C_1, and α. Examples of application to atmospheric observations by the founders of generalized scale invariance can be found in Chigirinskaya et al. (1994), Lazarev et al. (1994), and Schmitt et al. (1994). A more recent statement and application is Seuront et al. (1999).

Under the assumption of three-dimensional isotropy of turbulence and noting that the Navier–Stokes equation exhibits scale invariance, the velocity fluctuations, Δv, and temperature fluctuations, ΔT, are describable by the scaling relationships

$$\Delta v \approx \varepsilon^{1/3}\ell, \tag{4.1}$$

$$\Delta T \approx \phi^{1/3}\ell^{1/3}, \tag{4.2}$$

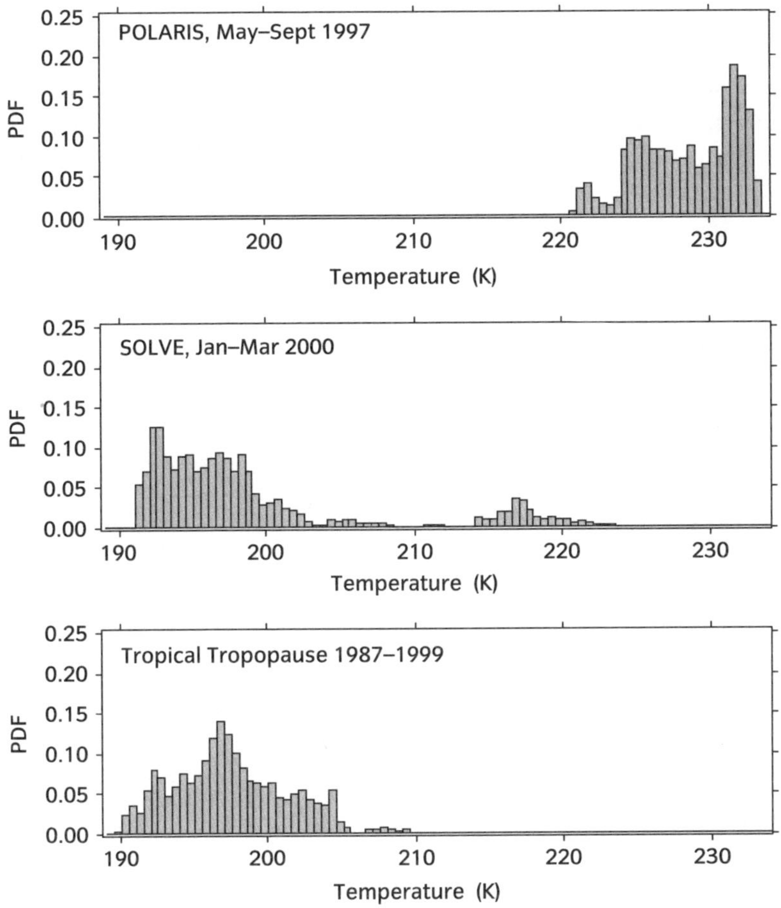

Figure 4.1 Ambient temperature PDFs from the Arctic summer of 1997, the Arctic winter of 2000, and the tropical tropopause, 1987–1999. The top panel shows data from the Arctic summer anticyclone in the lower stratosphere, which is subject to continuous radiative heating from solar photons. It has a skewed PDF, with a warm most probable value and a long, cool tail. The middle panel shows data from the Arctic lower stratospheric winter polar vortex, which is subject to infrared radiative cooling during the polar night. There is a cold most probable value with a long warm tail. The bottom panel refers to data taken near the tropical tropopause between 1987 and 1999. This region is also characterized by cold temperatures, like the winter vortex, but caused by adiabatic cooling in air ascending from the tropical troposphere. The distribution is again skewed, with a cold most probable value and a long warm tail. If these data are separated into individual missions in individual locations, the PDFs resemble the middle panel more, as does the PDF for the entire tropical tropopause from operational meteorological models. Again, these PDFs are characteristic of fractal behaviour, and point to the importance of turbulent heat flux on all scales.

where $\Delta v = |v(r + \ell) - v(r)|$ and $\Delta T = |T(r + \ell) - T(r)|$ are the velocity and temperature shears at length scale ℓ, ε is the dissipation rate of turbulent kinetic energy, and ϕ is the flux resulting from the nonlinear interaction of the v and T fields (temperature advection). ϕ is defined as $\varepsilon^{1/2}\chi^{3/2}$ where

χ is the rate of temperature variance flux. It is dimensional analysis applied to Richardson's idea of a cascade.

The observations show that this approach is insufficient; it fails to deal with intermittency. The basic idea is to treat intermittency as a set of singularities, whose distribution is characterized by the function $c(\gamma)$ according to the probability distribution

$$Pr(E_\lambda \geq \lambda^\gamma) \approx \lambda^{-c(\gamma)}. \tag{4.3}$$

Here λ is a scale ratio $= L/\ell$ where L is the outer scale, for example the circumference of the Earth. $c(\gamma)$ is a co-dimension, the difference between the dimension d of the physical, embedding space and the dimension $D(\gamma)$ describing the hierarchy of fractal dimensions associated with the different levels of singularity, that is the intermittency. In a space-filling turbulent field, for example, there would be λ^d possible vortices with $\lambda^{-D(\gamma)}$ vortices of different intensity, so

$$c(\gamma) = d - D(\gamma). \tag{4.4}$$

The probability distribution is expressible by its statistical moments. We introduce the scaling moment function $K(q)$ which describes the multiscaling of the statistical moments of order q:

$$\left\langle (E_\lambda)^q \right\rangle \approx \lambda^{K(q)} \tag{4.5}$$

and

$$K(q) = \max_\gamma \{q\gamma - c(\gamma)\}, \tag{4.6}$$

so

$$c(\gamma) = \max_q \{q\gamma - K(q)\}, \tag{4.7}$$

which implies a one-to-one correspondence between singularities γ and moments q.

The quantity H_1 is the scaling exponent calculated from an aircraft time series $f(t)$ by application of the first order structure function. The qth order structure function of $f(t)$ is defined by

$$S_q(r; f) = \langle |f(t + r) - f(t)|^q \rangle, \tag{4.8}$$

where the lag r is real and positive and the angle brackets denote an average over t and ensemble averaging over f. Note that $S_q(r; f)$ is a signal, it is not entropy.

If a plot of $\log S_q(r; f)$ versus $\log(r)$ is linear (the 95% confidence interval of the fit to obtain the slope $\zeta(q)$ is generally less than 10%), then $\zeta(q)$ is a scaling exponent for $f(t)$. We then define

$$H_q = \zeta(q)/q. \tag{4.9}$$

If H_q is constant as q changes, then $f(t)$ is monofractal and the scaling exponent $H = H_q$. If H_q is not constant with q, then $f(t)$ is multifractal and the scaling exponent is

$$H = H_q + K(q)/q. \tag{4.10}$$

where $K(q)$ is calculated during the estimation of intermittency.

Because $K(1) = 0$, the quantity H_1 is a good estimate of the scaling exponent in both monofractal and multifractal cases. The exponent C_1 measures the intermittency of the signal and takes on values from zero to unity. Values near zero characterize a signal with low intermittency, for example a Brownian motion, and values near unity characterize a signal which is highly intermittent, for example a Dirac δ-function. Values in the range 0.02 to 0.10 seem to characterize atmospheric quantities, and although they are towards the low end of the mathematically possible zero to unity range, they are significant in our context. Considering the signal $f(t)$ to have been observed at discrete time intervals $t = 1, 2, 3, \ldots, t_{\max}$, define

$$\varepsilon(1,t) = \frac{|f(t+1) - f(t)|}{\langle |f(t+1) - f(t)| \rangle}, \quad t = 1, 2, 3, \ldots t_{\max}, \tag{4.11}$$

$$\varepsilon(r,t) = \frac{1}{r} \sum_{j=t}^{t+r-1} \varepsilon(1,j), \quad t = 1, 2, 3, \ldots t_{\max} - r. \tag{4.12}$$

For our signals, it is found that the quantity $\langle \varepsilon(r,t)^q \rangle$ has a power law dependence on the scale r. An unweighted linear least squares fit to $\log\langle \varepsilon(r,t)^q \rangle$ versus $\log r$ provides a slope $-K(q)$. A plot of $K(q)$ versus q shows a convex function with $K(0) = K(1) = 0$. The exponent C_1 is defined as $K'(1)$, evaluated here numerically from the slope defined by the points $(0.9, K(0.9))$ and $(1.1, K(1.1))$. The uncertainty estimate in C_1 is obtained by taking the square root of the sum of the squares of the 95% confidence intervals returned by the unweighted linear least squares fits corresponding to $q = 0.9$ and $q = 1.1$.

The Lévy index α has the range $0 \leq \alpha \leq 2$ and measures the power law fall-off of the tail of the probability density of the signal increments. Gaussian noise or Brownian motion has $\alpha \approx 2$. Schertzer and Lovejoy (1991) discuss the five main cases for α; here we note that the variables we have measured appear to have $1 < \alpha < 2$. Our experience indicates that a large quantity of high quality data is necessary for an accurate computation of α. We use the double trace moment technique to compute α. Define

$$\varepsilon(1,\eta,t) = \frac{|f(t+1)-f(t)|^{\eta}}{\langle |f(t+1)-f(t)|^{\eta}\rangle}, \quad t = 1,2,3,\ldots t_{\max}, \tag{4.13}$$

$$\varepsilon(r,\eta,t) = \frac{1}{r}\sum_{j=t}^{t+r-1} \varepsilon(1,\eta,j), \quad t = 1,2,3,\ldots t_{\max}-r, \tag{4.14}$$

where η is allowed to range from -1.0 to $+1.0$ in steps of 0.1. For $q = 1.5$ an unweighted fit to $\log\langle\varepsilon(r,\eta,t)^q\rangle$ versus $\log r$ is made, with the slope being $K(q,\eta)$ and the standard deviation being $\sigma(q,\eta)$. A plot of $\log K(q,\eta)$ versus $\log \eta$ yields for our data a collinear region having a positive slope. A weighted linear least squares fit to this region, with weights $K(q,\eta)\ln 10/\sigma(q,\eta)$, has slope α, with the uncertainty represented by the 95% confidence interval returned by the weighted fit. Further discussion is available (Tuck et al. 2002).

The relationships of $K(q)$ to intermittency C_1 and Lévy exponent α were obtained by Schertzer and Lovejoy from study of the limiting behaviour of λ in the limits of unity and infinity:

$$K(q) = \begin{cases} \dfrac{C_1}{\alpha-1}(q^{\alpha}-q), & \alpha \neq 1 \\ C_1 q \ln q, & \alpha = 1 \end{cases} \tag{4.15}$$

and

$$c(\gamma) = \begin{cases} C_1\left(\dfrac{\gamma}{C_1\alpha'}+\dfrac{1}{\alpha}\right)^{\alpha'}, & \alpha \neq 1 \\ C_1 \exp\left(\dfrac{\gamma}{C_1}-1\right), & \alpha = 1 \end{cases}, \tag{4.16}$$

where

$$\frac{1}{\alpha}+\frac{1}{\alpha'} = 1 \tag{4.17}$$

and $\alpha' = \alpha = 2$ is the Gaussian; when $\alpha' > 2$, $1 < \alpha < 2$; when $\alpha' < 0$, $0 < \alpha < 1$.

In the next section, we consider the results of calculating H_1, the more robust exponent, from a variety of aircraft data.

4.2 Scaling of observations: H_1

In actual practice, the scaling exponent H_1 is more robust to missing data than are C_1 and α, which can be rendered inaccurate by even a few missing data points or noise spikes arising from, for example, known instrumental noise sources such as cosmic ray impacts on detectors, plasma instabilities in light sources, or telemetry errors in the case of dropsondes. There are many more suitable flight segments that can sustain analysis for H_1 than there are for the intermittency and for the Lévy exponent, both of which are sensitive to a few errors because they pertain more to the infrequent, relatively high amplitude events constituting the long tail of the PDFs. The same is true of the different variables—the instrumentation for wind, temperature, and pressure has been long established, while some of the chemical instruments measure molecules which are present at extremely low mixing ratios and do not have the data continuity and signal-to-noise that would enable a full generalized scale invariance analysis. Nevertheless, with a few rare exceptions, it is apparent that atmospheric variability dominates over instrumental noise. This is evident in the PDFs, which only show Gaussian or Poisson distributions at the very short scales where random instrumental noise dominates; indeed, the log-log plots forming the variograms are excellent, direct diagnostics of this situation. When it occurs, the slope flattens at the smallest scales, so $H_1 \rightarrow 0$.

We first examine composite variograms from 'horizontal' flight legs of four different aircraft: the ER-2, which is entirely lower stratospheric; the WB57F at low latitudes (between 5°N and 33°N), which is upper tropospheric and lower stratospheric; the DC-8 at high southern latitudes, which is largely upper tropospheric; and the G4, which is largely upper tropospheric over the eastern Pacific Ocean from 10°N to 60°N. Composite variograms are a way of using all the data to obtain a representative or canonical scaling exponent H_1 with very low error bars. A variogram is a means of describing positional relationships between two points in a data set; it is possible to have two time series, for example, with nearly identical single-point statistics such as means, standard deviations and medians yet which have differing 'textures' or 'roughnesses'. The two-point statistic is plotted in log-log form for all possible separation distances between all possible pairs of points. We shall refer to the first moment version of such a plot, like Figures 2.6 and 2.7, as a variogram, a usage encompassed by the wider definitions of the word to be found in the literature; note, however, that a variogram is often given a narrower definition in which the average (half) squared separation between the measured values at different locations is used. Such a narrow definition is associated with traditional Gaussian assumptions, second moments, variances, power spectra, and the like. Our approach, as defined in Equation (4.8) with $q = 1$, follows the wider definition. The ER-2 data are shown in Figure 4.2, the WB57F data are shown in Figure 4.3, the DC-8 data in Figure 4.4, and the G4 data in Figure 4.5. Both wind speed and temperature show values of H_1 close to the theoretical value of 5/9. We note that unlike wind speed and temperature, nitrous oxide, a passive scalar (tracer) at about 300 ppbv mixing ratio, can have no effect on the motion of the aeroplane. The lower

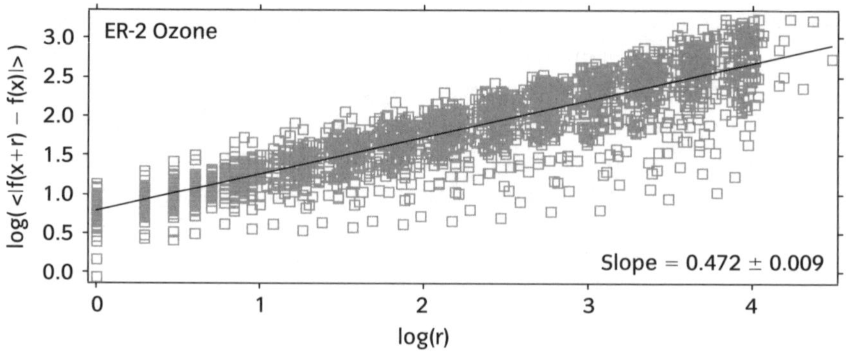

Figure 4.2 As Figure 2.6 but for ozone. The mean, which reflects all ER-2 'horizontal' flight segments of over 2000 seconds under autopilot control between 1987 and 2000, indicates that on average ozone does not behave like a passive scalar in the lower stratosphere, having H_1 of 0.47 rather than the 0.56 characteristic of a passive scalar. However, the upper part of the envelope suggests that there are some regimes where ozone is a tracer, while the lower envelope is defined by flights in air where the photochemistry was sufficiently rapid to yield scaling exponents H_1 in the region of 0.25 to 0.33, inside the polar vortices. Even in the presence of a sink, the ozone still shows scaling behaviour.

stratospheric ozone data and the upper tropospheric humidity data show deviations from this value, which we attribute to the fact that they are not passive scalars—ozone photochemistry and precipitation respectively constitute source/sink processes—it is not possible a priori to tell which. The attribution to geophysical phenomena and not to instrumental artefacts is justified by the examination of the effects of instrumental noise in Tuck et al. (2003b) and at the end of Chapter 1. There is a set of observations of a known passive scalar, nitrous oxide, which is good enough to sustain the statistical multifractal analysis; Figure 4.6 shows its H_1 to be in excellent agreement with the theoretical value of 5/9. Taken together, the analyses of these observational data from the different aircraft establishes generalized scale invariance in the upper troposphere and lower stratosphere over a wide range of locations. One point to note is that aircraft, when flying under autopilot control at some steady condition such as pressure altitude or Mach number, themselves respond to the turbulent structure of the wind and temperature fields on a wide range of scales, depending upon aircraft speed and inertia (Lovejoy et al. 2004) and the autopilot control algorithm responses, which are generally not recorded. At the smaller scales the aircraft cannot respond to the fluctuations, whereas at the longest scales the constant corrections made to navigate to a particular point tend to produce a flight track resembling the smooth continuous lines seen, for example, in flight routes portrayed in the magazines of commercial airlines. The result is a flattening of the log-log plot at small scales and a steepening of it at large ones, with a nett small effect because of a tendency for these effects to

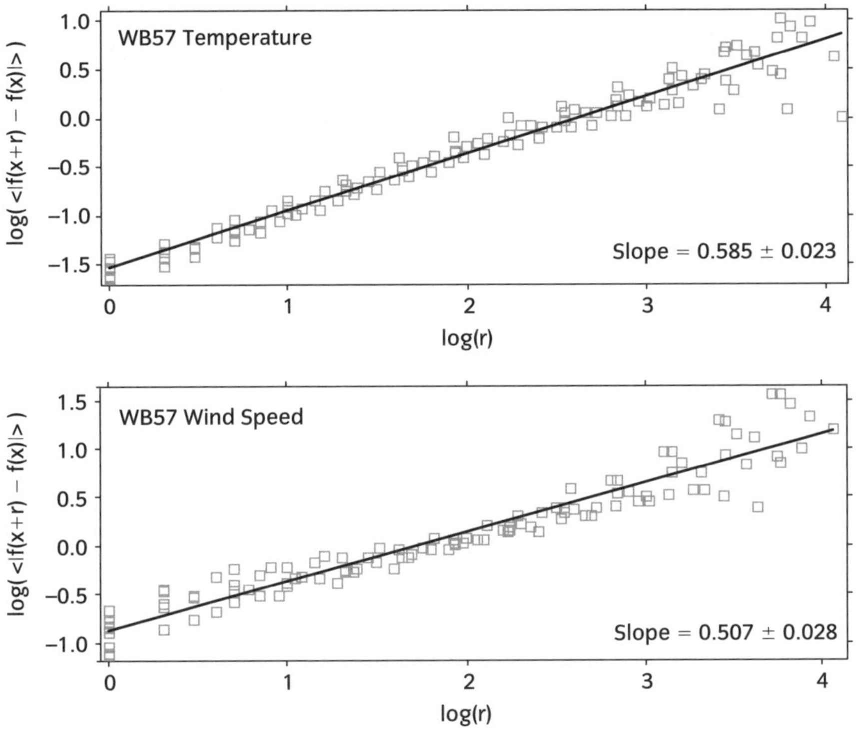

Figure 4.3 Composite variograms for temperature and wind speed measured along level WB57F flight legs near the tropical tropopause. The observations were taken in 1998 and 1999 during the WAM and ACCENT missions, based mainly at Ellington Field (30°N, 95°W) with some excursions to San José, Costa Rica (10°N, 84°W). The wind speed data are from the aircraft navigation system, not research instrumentation. The slopes yield scaling exponents H_1 close to the value of 5/9 predicted by general scale invariance, given that the aircraft samples vertical motion even in 'horizontal' flight.

cancel in the calculation of the slope, which yields H_1. Comparison of H_1 values between different variables and different atmospheric régimes therefore can and does produce consistent behaviour and interpretations, as we will see in succeeding pages of this chapter.

It is also possible to apply generalized scale invariance to the vertical scaling of the entire troposphere, by means of sensors for temperature, pressure, and relative humidity dropped by parachute from an aircraft, which also tracks these dropsondes by GPS, from which the velocities of both the sonde and the atmosphere, that is the wind, can be calculated (Hock and Franklin 1999). The G4 flights shown in Figure 4.5 were associated with the dropping of 261 GPS sondes over the eastern Pacific Ocean in February–March 2004 during the NOAA Winter Storms project, which recorded at 2 Hz horizontal wind speed, temperature, pressure, and relative humidity during the ~800 seconds it took to fall from about 13 km. The composite variograms describing the vertical scaling H_1 of the three variables are shown

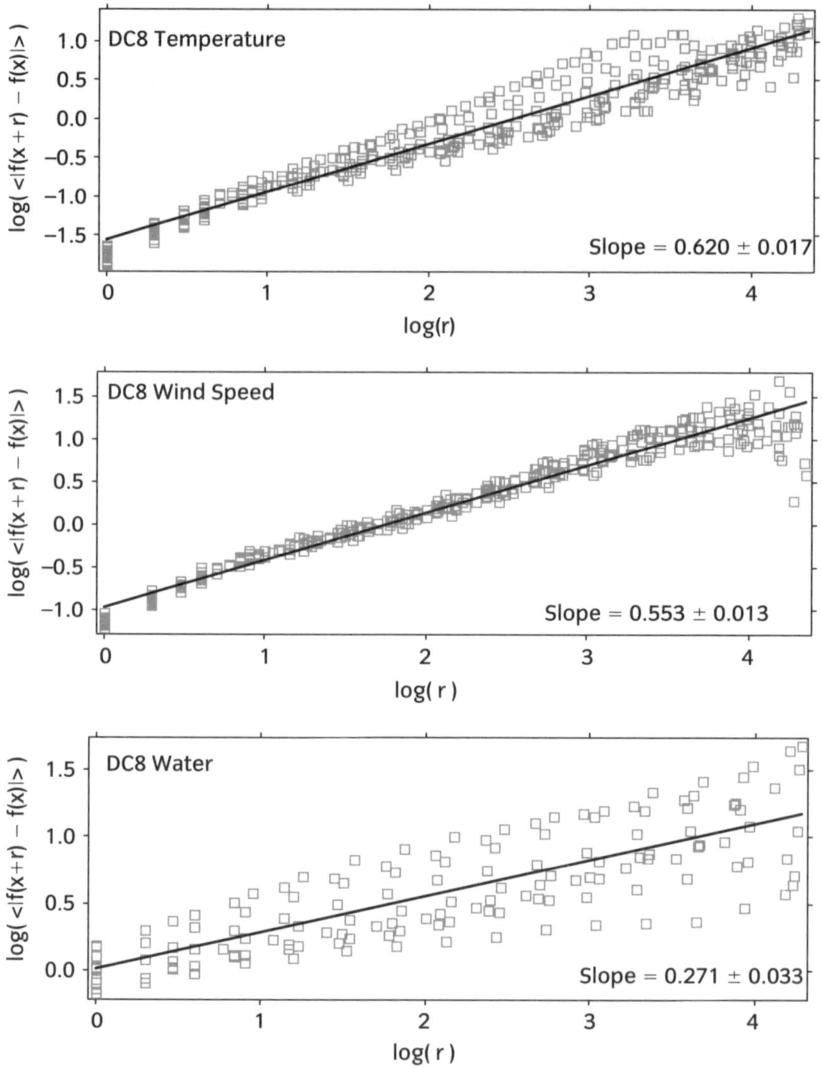

Figure 4.4 Composite variograms for temperature, wind speed, and water measured along level DC-8 flight legs during AAOE at 1 Hz, out of Punta Arenas (53°S, 71°W) in August to September 1987. The flights were over west Antarctica and ranged as far south as the pole, mainly in the upper troposphere with some segments in the lowermost stratosphere. The temperature and wind speed data are not of the same quality as the ER-2 observations and may have some intervals of manual rather than autopilot control, but nevertheless again lend support for the predictions of generalized scale invariance. The total water data represent the total of vapour and ice measured as vapour (Kelly et al. 1991) and show that in the upper troposphere and lowermost stratosphere over Antarctica in late winter and early spring that water substance is on average not a passive scalar; ice crystals sedimenting under gravity are therefore acting as a sink for total water, although one flight segment does have $H_1 = 0.53$. Note that even with a sink operative, the total water data still show the scaling behaviour characteristic of statistical multifractality.

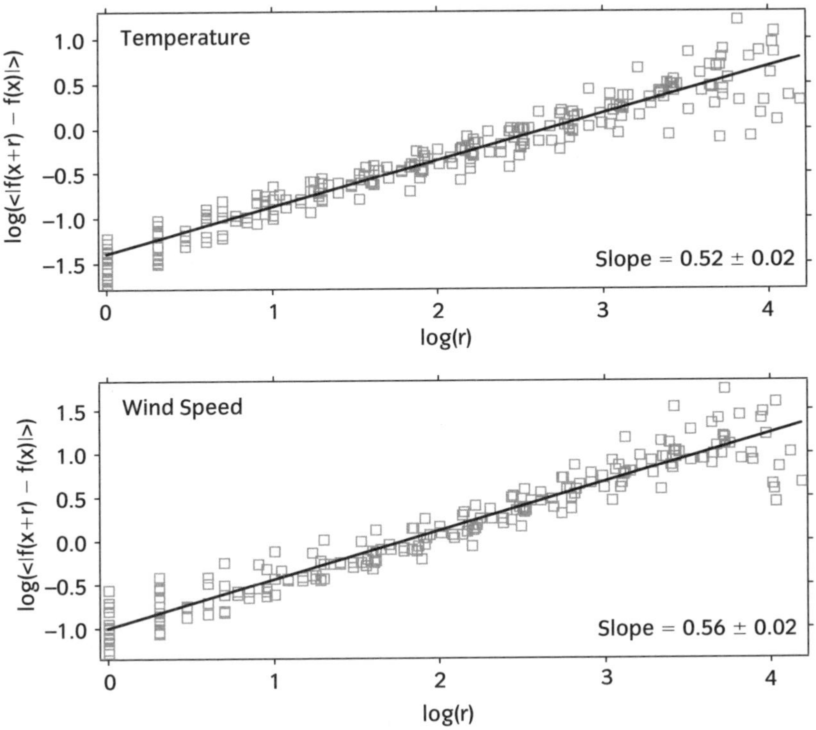

Figure 4.5 Composite variograms for temperature and wind speed measured along level Gulfstream 4SP flight legs during Winter Storms 2004. These data were taken, largely in the upper troposphere, over the eastern Pacific Ocean between 15°N and 60°N between late February and mid March in 2004. The scaling behaviour seen 'horizontally' in the lower stratosphere by the ER-2 is extended by these results into the upper troposphere.

in Figure 4.7. All the H_1 exponents from the horizontal aircraft data, the 'vertical' data from the aircraft and the vertical data from the dropsondes are summarized in Table 4.1.

According to the theory of generalized scale invariance (Schertzer and Lovejoy 1985, 1987, 1991), the ratio of the horizontal to vertical conservation scaling exponents H is 5/9, consisting of the ratio 1/3 ÷ 3/5. The 1/3 in the horizontal originated with Kolmogorov (1962, 1991) and the 3/5 in the vertical with Bolgiano (1959). The vertical exponents are greater than 3/5 for both dropsonde and aircraft data, being near but significantly less than unity for temperature and about 0.75 and 0.68 for both relative humidity and wind speed (Table 4.1). The horizontal scaling exponents for the aircraft data are 0.52, 0.45, and 0.56 respectively for temperature, relative humidity, and wind speed. For the aircraft data, the ratios of the horizontal exponents to the vertical ones are 0.55 (temperature), 0.67 (relative humidity), and 0.82 (horizontal wind speed). The generalized scale invariance analysis of the vertical structure of the horizontal wind speed by

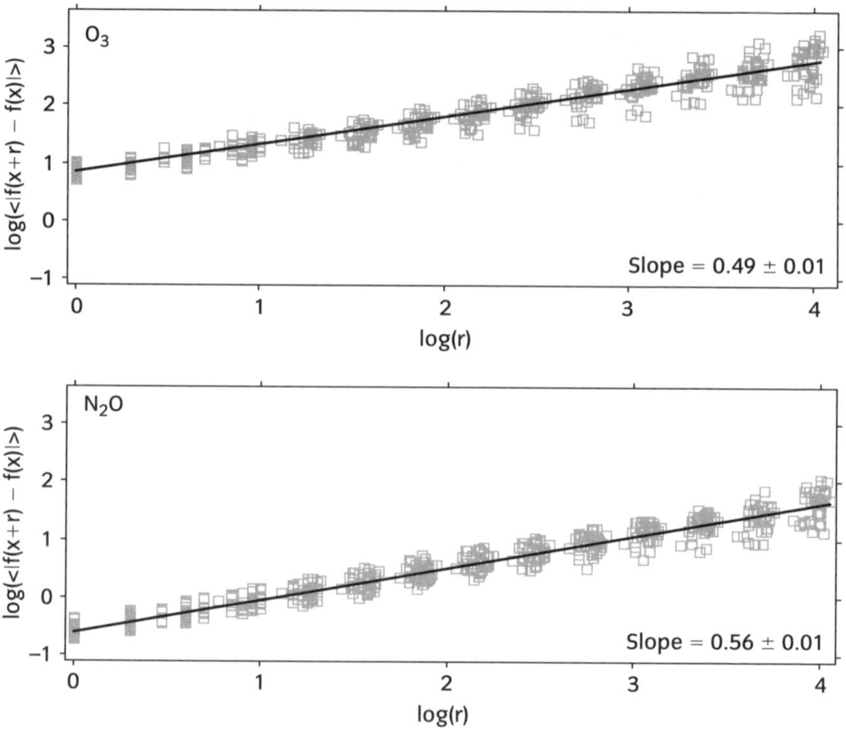

Figure 4.6 Composite variograms for ozone (Proffitt et al. 1989) and nitrous oxide (Loewenstein et al. 1989) measured along level ER-2 flight legs during ASHOE/MAESA. These data were acquired between February and November 1994, ranging between 59°N and 70°S with a heavy weighting towards the Southern Hemisphere and for time outside the polar vortices. Nitrous oxide, known to be a tracer (passive scalar) in the lower stratosphere, closely obeys the $H_1 = 5/9$ predicted by general scale invariance; its value cannot affect the motion of the aircraft, unlike the wind speed and temperature. The ozone shows the presence of a sink, presumably chemical, with $H_1 = 0.49$. There is no theory yet linking the value of H_1 with the rate of the sink, but comparison with the polar vortex part of the data in Figure 4.2 suggests that there is an empirical correlation.

Lazarev et al. (1994) found $H = 0.50 \pm 0.05$ from the ascent phase of 287 radiosonde launches, truncated to include only data up to 13 km altitude. This contrasts with our value from 235 dropsondes of 0.768 ± 0.005 for an almost identical altitude range. It is known that balloon payloads are in the turbulent wake of the balloon on ascent from the behaviour of the water vapor sensor, even when attached on long cables, so that data are completely reliable only on descent. This is less true of ozone and temperature, but may have affected the analysis of earlier stratospheric large balloon data, for example Tuck et al. (1999); we therefore suggest that the dropsonde data are more credible. The behaviour of upper tropospheric aircraft observations of humidity has been discussed previously, with the conclusion that in both hemispheres air was dried by the cooling during sloping motion

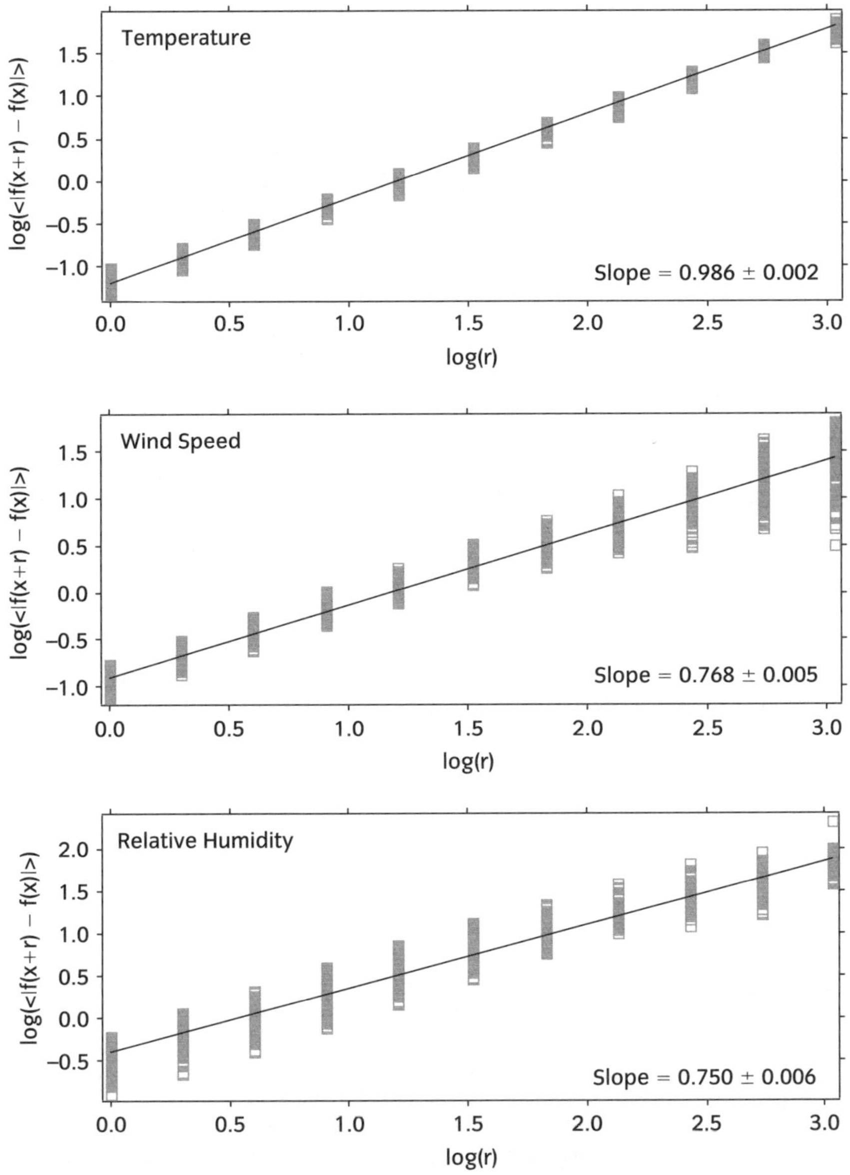

Figure 4.7 Composite variograms for temperature, wind speed, and relative humidity measured by all drop sondes in the Winter Storms 2004 mission; the sondes were dropped during the 'horizontal' flight segments shown in Figure 4.5. These observations allow examination of the scaling behaviour in the 'vertical' between 13 km and the surface. The values of H_1 for temperature and wind speed are not the 3/5 predicted by Bolgiano-Obukhov theory, but scaling is respected. There is discussion of this result in the text. The near but significantly less than unity value for temperature reflects the importance of gravity acting through hydrostatic balance; if $H_1(T)$ is evaluated via spectral analysis rather than first moments, the value is 1.25, which is still anomalous compared to the 0.7–0.8 range for horizontal wind speed and relative humidity obtained by either method.

Table 4.1 Scaling exponents H_1 from dropsondes and aircraft Winter Storms 2004.

	Dropsondes	Vertical Aircraft Segments	Horizontal Aircraft Segments	Ratio Vert/Horiz
Temperature	0.986 ± 0.002	0.95 ± 0.02	0.52 ± 0.02	0.55 ± 0.02
Wind Speed	0.768 ± 0.005	0.68 ± 0.02	0.56 ± 0.02	0.82 ± 0.01
Relative Humidity	0.750 ± 0.006	0.66 ± 0.03	0.45 ± 0.03	0.68 ± 0.01

up isentropes to the west of anticyclones (ridges) in the subpolar regions and to their east in the subtropics (Kelly et al. 1991; Yang and Pierrehumbert 1994); scaling behaviour indicated that water was not a passive scalar (conservative tracer) in the subtropical case (Tuck et al. 2003b). Galewsky et al. (2005) have confirmed this drying mechanism globally in the subtropics via study of both re-analysis and general circulation model fields. The departure from passive scalar behaviour for humidity is not therefore surprising.

The vertical scaling of temperature from the dropsondes ($1 > H > 0.95$) appears to indicate that gravity, as expressed in the hydrostatic Equation (4.18), has a major influence on the structure of temperature, almost but not quite to the point of obviating fractality to give the conservative behaviour (perfect neighbour-to-neighbour correlation for all intervals) which would be associated with $H = 1$.

$$\frac{\partial p}{\partial z} = -\rho g \tag{4.18}$$

where p is pressure, z is the altitude, ρ is density, and g is acceleration due to gravity.

The vertical exponents are significantly less for relative humidity and wind speed, but still larger than theoretical expectation from generalized scale invariance, although scaling is respected. The aircraft ascents and descents have some small horizontal components, and while this is reflected in the slightly ($\leq 10\%$) lower vertical exponents for all three variables, they basically confirm the dropsonde results, as do independent WB57F profiles taken through the depth of the tropical troposphere during pre-AVE (Aura Validation Experiment) (Fig. 4.8). The perspective offered here on the effect of gravity on the vertical scaling of temperature is of interest in view of the recent discussion of the hydrostatic approximation in numerical weather prediction (White 2002; Davies et al. 2005; White et al. 2005).

There is unexpected structure hidden in some of the canonical H_1 exponents obtained from the variograms. This is particularly true of the horizontal and vertical wind speeds associated with jet streams, and of the horizontal temperature gradient in the cases of the stratospheric polar night jet stream and the subtropical jet stream. When the $H_1(s)$, where s is wind

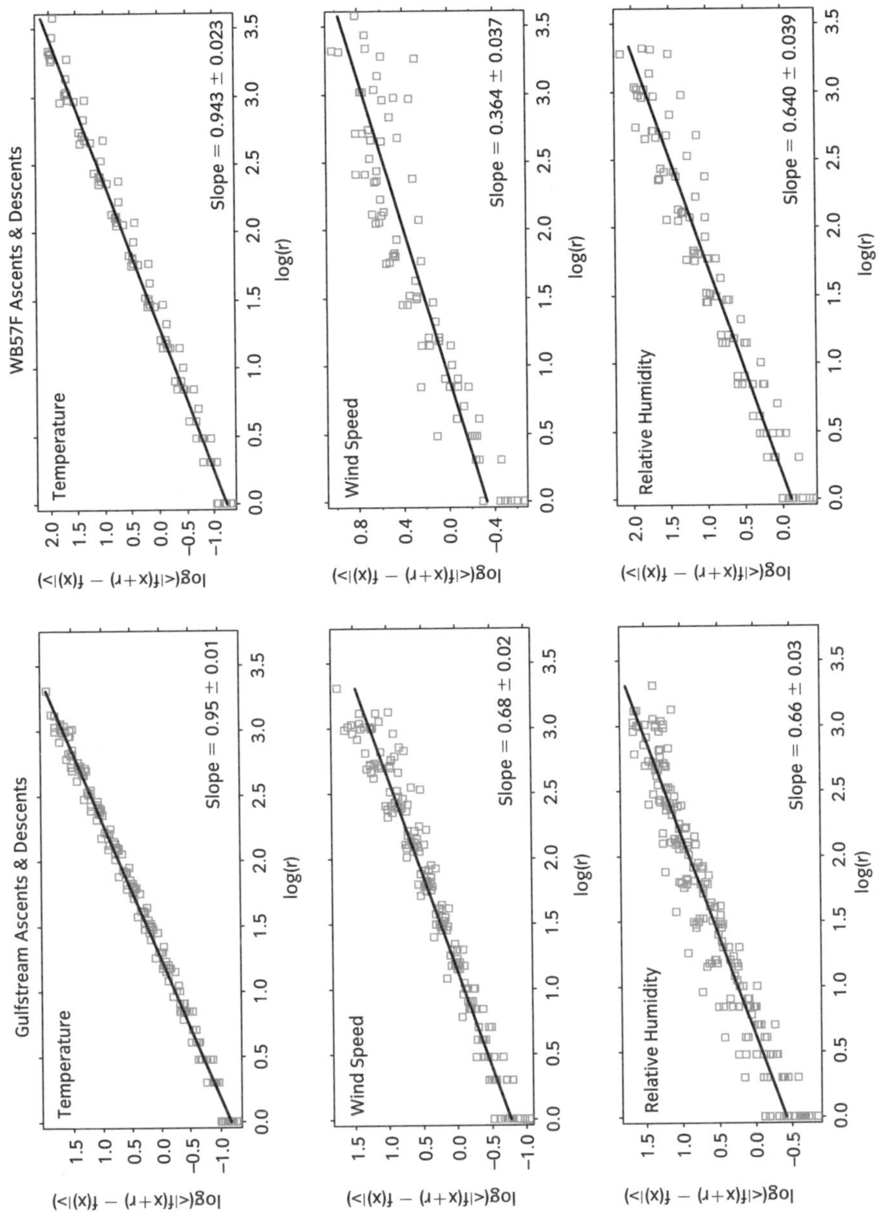

Figure 4.8 Composite variograms for temperature, wind speed, and relative humidity for the ascent and descents of the Gulfstream 4SP (16 near Honolulu, 2 near Anchorage and 2 near Long Beach) during Winter Storms 2004 and for the 16 ascents and descents of the WB57F over San José, Costa Rica, during pre-AVE. The WB57F wind speed data are not from research instrumentation and are not as accurate as the temperature and relative humidity. All the flights took place in the first three months of 2004. There is a horizontal component in these profiles; the vertical component averages about 14 ms^{-1} while the horizontal one is an order of magnitude larger. The vertical component is nearly the same as the fall speed of the dropsondes whose data are shown in Figure 4.7. The scaling is similar to that of the dropsondes, but with a small decrease arising from the horizontal component in the aircraft motion; the aircraft profiles can therefore be thought of as 'vertical'.

speed, from individual horizontal flight segments are plotted against the horizontal wind shear, a correlation of $H_1(s)$ with the strength of the jet stream is seen, see Figure 4.9(a). The same is true for $H_1(T)$ and the horizontal temperature gradient. An unforced, untheoretical link with large-scale meteorology has emerged. This result for $H_1(s)$ is not confined to horizontal data from the ER-2 in the lower stratosphere. It is evident vertically in the tropospheric dropsonde data over the eastern Pacific Ocean, for the subtropical and polar front jet streams as well as the stratospheric polar night jet stream, see Figure 4.10. This could be seen as long-range correlation or large-scale order, emerging from the operation of the fluctuation-dissipation theorem and maximization of entropy production, via the upscale propagation of enstrophy, or half the squared vorticity. Jet streams are the product

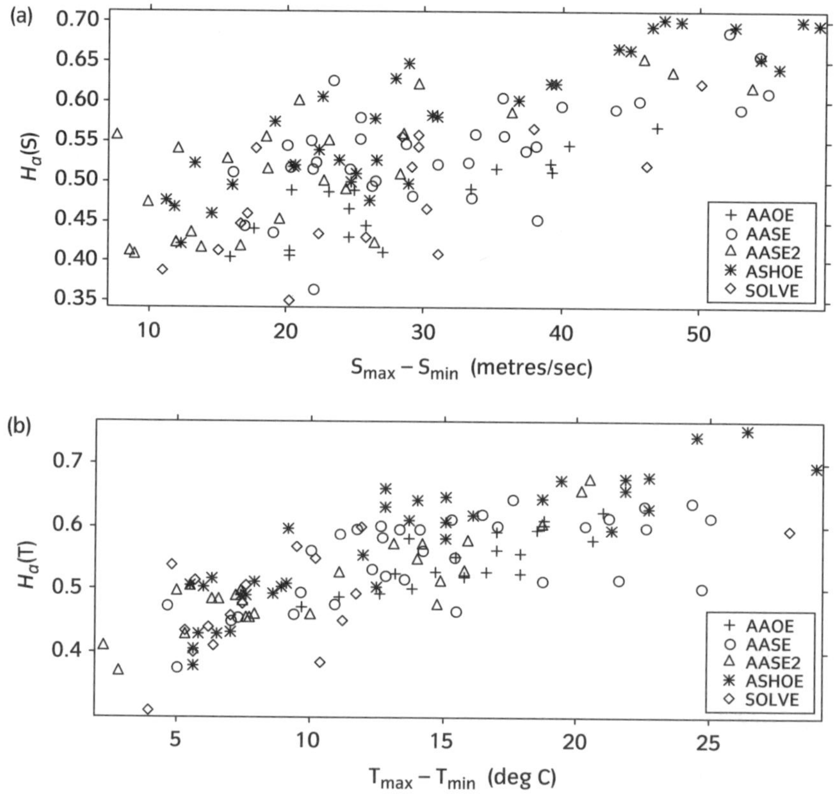

Figure 4.9 Upper, scaling exponent H_1 versus wind speed range (a measure of the horizontal wind shear), and, lower, versus temperature range (a measure of the horizontal temperature gradient), for the 'horizontal' ER-2 flight segments during the AAOE, AASE, AASE2, ASHOE, and SOLVE missions. Note that a correlation has emerged between the scaling exponents and the magnitudes of the large-scale gradients of wind speed and of temperature. There is an observational link between the scaling arising from treating the smaller scales and the strength of the jet stream flow, a traditional entity in the large-scale framework that is dynamical meteorology.

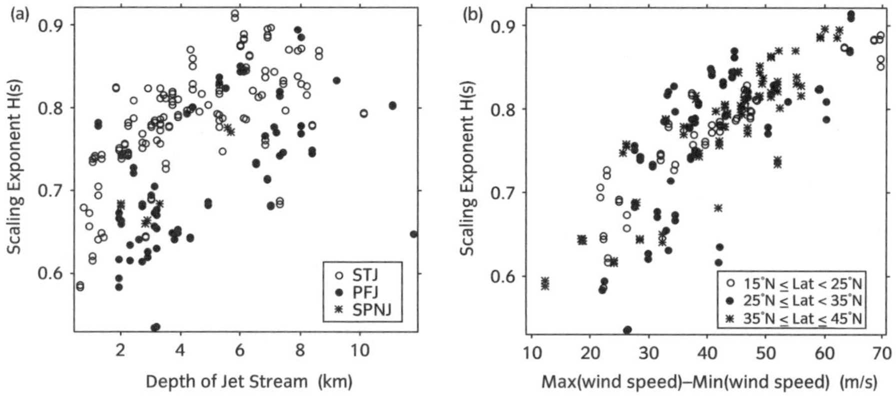

Figure 4.10 Left, the vertical scaling exponent H_1 for horizontal wind speed s as a function of jet stream depth, obtained from dropsonde data. The appellations STJ, PFJ, and SPNJ denote respectively the sub tropical, polar front, and stratospheric polar night jet streams. The STJ seems to have somewhat higher values of $H_1(s)$ for a given jet stream depth. Right, $H_1(s)$ versus the difference between the maximum and minimum wind speeds recorded on the sounding, a measure of the vertical shear of the horizontal wind. As shown in the horizontal in the upper panel of Figure 4.9, the scaling of the horizontal wind speed in the vertical is correlated with vertical measures of jet stream strength. The emergent link from the scaling analysis of the smaller scale fluctuations with larger scale meteorological concepts exists in the vertical, too.

of these processes interacting with the planetary scale boundary conditions, the planetary rotation, gravity, and the solar beam. The mixing effect of the speed and directional wind shears acting on air from different regions entering and leaving the jet is important, as has long been realized (Rossby 1947; Riehl 1954; Reiter 1963). There has been a recent renewal of interest from the standpoint of generalized scale invariance (Tuck et al. 2004) and from the combination of 'PV thinking' (Hoskins et al. 1985) and 'standard' turbulence theory, see Baldwin et al. (2007). As we have noted elsewhere, the core speeds can be more than 30% of the most probable molecular velocity, a circumstance strengthening the need for a molecular perspective on turbulence and the generation and propagation of vorticity. It also points to the possibility of the upscale propagation of high molecular speeds and the associated vorticity structures in jet streams, via the positive feedbacks associated with 'ring currents'.

There is some observational evidence that vorticity and potential vorticity from the polar reservoir in the winter lower stratosphere, as embodied in the lower stratospheric vortex, tends to accumulate on the cyclonic (poleward) flank of the subtropical jet stream, particularly over the continents (Tuck et al. 1992; Tuck 1993; Tripathi et al. 2006). Similarly, low vorticity and potential vorticity filaments are found wrapped around the polar night jet stream that forms the vortex (Tuck et al. 1997). Both conditions suggest that the exchange of contrasting vorticity structures between jet streams and their environment is a common process. The results are compatible with upscale propagation of vorticity structures, shaped by the land–sea distribution and carried by a positive feedback between long-tailed velocity distributions in molecules, turbulent vorticity structures, and the jet stream core, to maintain the jet stream itself.

There was no correlation of the vertical scaling exponents $H(T)$ and $H(RH)$ from the dropsonde data with either the depth of the jet streams or with the magnitude of the vertical shears of the horizontal wind speed. There was however, significant correlation of the vertical, dropsonde-derived $H(s)$ with these quantities for all three categories of jet stream, see Figure 4.10a for the correlation with depth of the jet stream. There is some separation of the subtropical jet stream exponents from those of the polar front and stratospheric polar night jet streams on this plot. Figure 4.10b shows a cleaner correlation of $H(s)$ with the vertical shear of the horizontal wind, echoing that found in the horizontal for the stratospheric polar night jet stream (Tuck et al. 2004), with little evidence of separation among the three latitude bands.

The correlation of $H(s)$ with jet stream depth, and the concomitant correlation with the vertical shear of the horizontal wind, underlines the remarks in Gill (1982) and White (2002) on the dynamics of wind in the presence of hydrostatic balance, especially the quantitative aspects. The fact that the vertical scaling of the horizontal wind in the troposphere and lowermost stratosphere yields an observational link between the small scale turbulent structure and the large scale manifestation of jet streams echoes the behaviour reported earlier for the horizontal wind speed of the SPNJ (Stratospheric Polar Night Jet) higher up in the stratosphere (Tuck et al. 2004). Both results demonstrate the utility of generalized scale invariance (Schertzer and Lovejoy 1985, 1987, 1991) in bringing coherence to an otherwise unruly set of high resolution, precise, and accurate airborne and dropsonde data.

While scaling is still respected in the vertical, for both dropsonde and aircraft data, we have no explanation for the observed deviations of the numerical values of the exponents and their ratios from those predicted by generalized scale invariance, which has hitherto between successful in accounting for the scaling behaviour observed in 'horizontal' flight in the lower stratosphere (Lovejoy et al. 2001, 2004; Tuck et al. 2002, 2003a, b, 2004, 2005).

Two further conceptual conclusions have been drawn from H_1 analyses of the high-resolution vertical data from the dropsondes. One of these is that apparently stable layers in the atmosphere actually have embedded unstable layers, each of which in turn has embedded stable layers and so on in a 'Russian doll' structure which constitutes a fractal set (Lovejoy et al. 2007b). This structure is not evident when the vertical resolution is degraded from 5 m to 500 m, as may be seen from Figure 4.11. In general

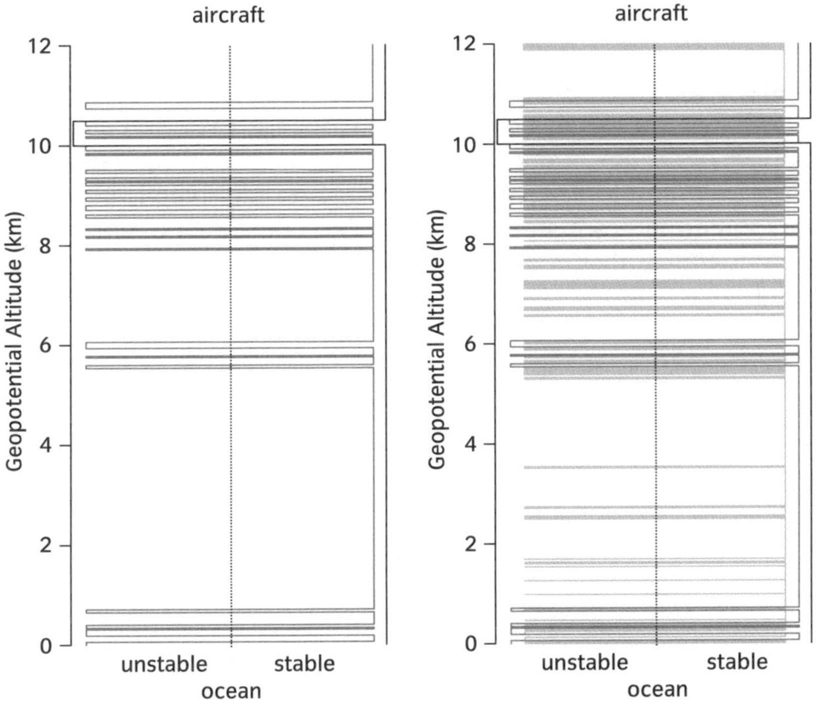

Figure 4.11 Left, the dynamical stability of the atmosphere, determined from a dropsonde (#2) from the NOAA Gulfstream 4SP aircraft at about 13 km on 20040229 at (25° N, 157° W), using the criterion $Ri > 1/4$. The result using vertical resolutions of 500 m and 50 m reveals a stable upper troposphere and an unstable lower troposphere for the former, and a few embedded layers of opposite stability in each for the latter. Right, the same procedure using vertical resolutions of 500 m, 50 m and 15 m. There are no unstable layers at scales smaller than about 80 m, but the stable layers are evident on all scales. At their high resolutions, using successive factors of 4 in resolution, a Cantor-like set with a 'Russian doll' structure is revealed, having a fractal dimension for the unstable layers of 0.65 ± 0.02, corresponding to a correlation dimension of 0.35 (Lovejoy et al. 2007b).

the upper troposphere is stable and the lower troposphere is unstable under this analysis at the coarsest vertical resolution, in accordance with meteorological expectation of the Pacific Ocean in late winter–early spring. The Richardson number Ri was examined, where

$$Ri = \frac{(g/\theta)(\partial\theta/\partial z)}{(\partial u/\partial z)^2} \tag{4.19}$$

In this equation, g is gravity, θ is potential temperature, u is the horizontal wind speed, and z is altitude. It expresses the energetic competition between buoyancy forces and the vertical shear of the horizontal wind; $Ri < 1/4$ is a necessary but not sufficient condition for instability. The insufficiency stems from the fact that even if there is sufficient energy in the wind shear for instability, there is no guarantee that it will be so utilized by the buoyancy forces. One view of this is that if the molecular speed distribution in a particular unstable layer embedded in a stable layer has a sufficiently over-populated high speed tail, it will be able to propagate vorticity structures up to the scale of the host stable layer in which it is embedded, so destabilizing it. If not, it will not be so enabled. Richardson instability, then, is a clue to how important such small-scale structure and process is in the atmosphere's continual thermodynamic drive to dissipate its energy, ultimately into molecular velocity.

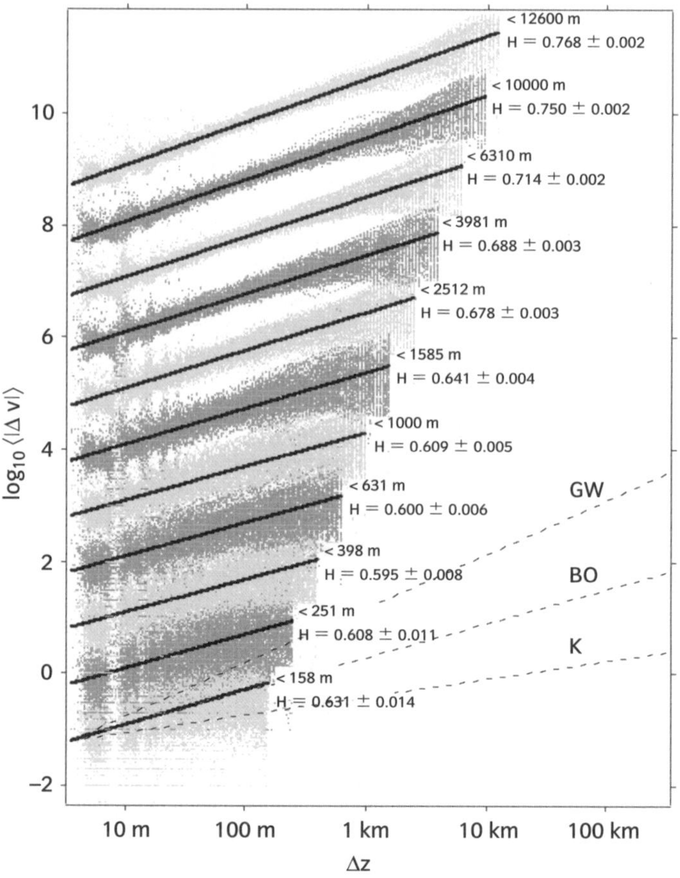

Figure 4.12 The vertical scaling of the horizontal wind, from all 235 useable dropsondes from the NOAA Gulfstream 4SP over the eastern Pacific Ocean during the Winter Storms 2004 project, February–March. The fits are root-mean-square to the vertical shears across layers of thickness increasing logarithmically. The reference lines have slopes corresponding to $H_1 = 1$ (gravity waves), $H_1 = 3/5$ (buoyancy, Bolgiano-Obukhov), and $H_1 = 1/3$ (Kolmogorov). Bolgiano-Obukhov is a good fit in the lower troposphere, but jet streams cause a systematic increase to about 0.8 when the upper parts of the profiles are included. In any event, isotropy is excluded. See text for further discussion. (Lovejoy et al. 2007a).

A second conceptual result from the H_1 analysis of the dropsonde data is that anisotropy is sufficiently influential in the atmosphere that isotropic turbulence is not observed, on any scale (Lovejoy et al. 2007a). This may be concluded from the vertical scaling of the horizontal wind, shown in Figure 4.12. Isotropic, Kolmogorov, scaling would result in $H_1 = 1/3$, with $H_1 = 3/5$ for Bolgiano scaling, which incorporates buoyancy, and $H_1 = 1$ for a linear gravity wave approach. It is clear the Bolgiano scaling is the closest to observation; the increase to values ~ 0.8 in the upper regions of the 13 km deep profiles is correlated with the presence of jet streams, a result discussed earlier in this section. The central point is that the observed horizontal and vertical scaling exponents are sufficiently different that isotropic turbulence is eliminated as a possibility on any scale.

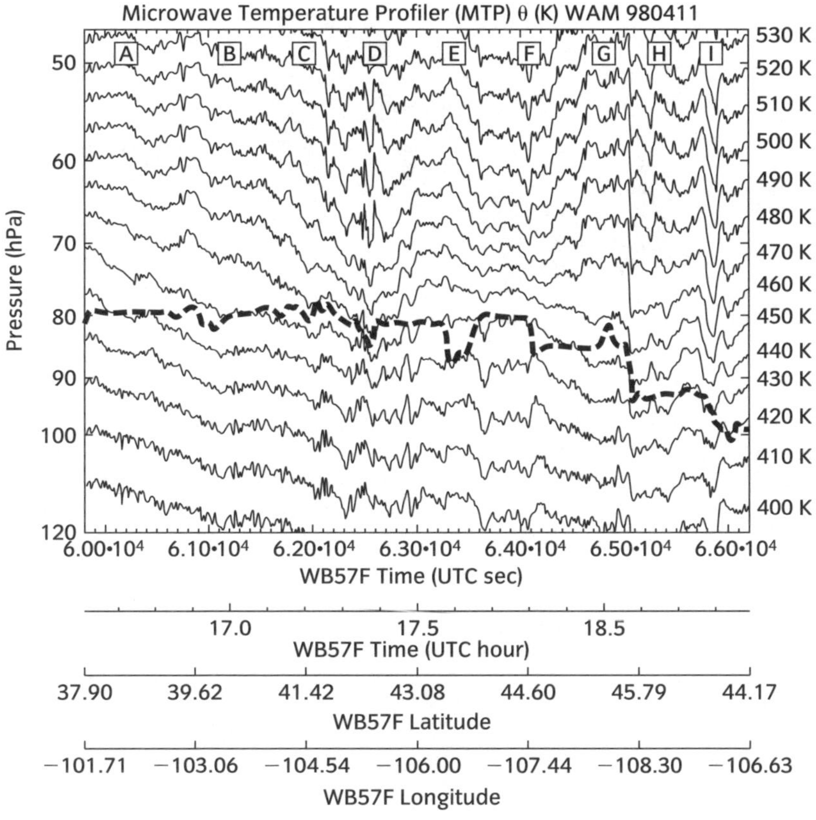

Figure 4.13 Potential temperature surfaces (K) from Microwave Temperature Profiler (MTP) on board the WB57F on April 11, 1998 corresponding to a lower stratospheric flight section from Texas to Montana. Contour interval is 10 K. Letters 'A' through 'I' correspond to events in the vertical structure of the isentropes. Flight altitude is plotted as dashed curve. Major gravity wave events are located at 'D' and 'I'. The near vertical isentropes at 'D' and 'I' show the possibility of vertical mass transport, particularly given the turbulent wind trace in Figure 4.14. The visible variability is essentially atmospheric (Gary et al. 1989; Gary 2006).

Because it offers a comparison with a high resolution 'forecast' by a numerical model, we next consider briefly an example of a high altitude flight by the WB57F over the Rocky Mountains, during which severe turbulence was encountered. Figure 4.13 shows the 'curtain' of isentropes along the flight track, obtained by the microwave temperature profiler instrument; the aircraft flight track is shown by the heavy dashed trace at about 80–90 hPa. The traces of temperature, wind speed, and wind direction derived from the aircraft are shown in Figure 4.14. The variability in the

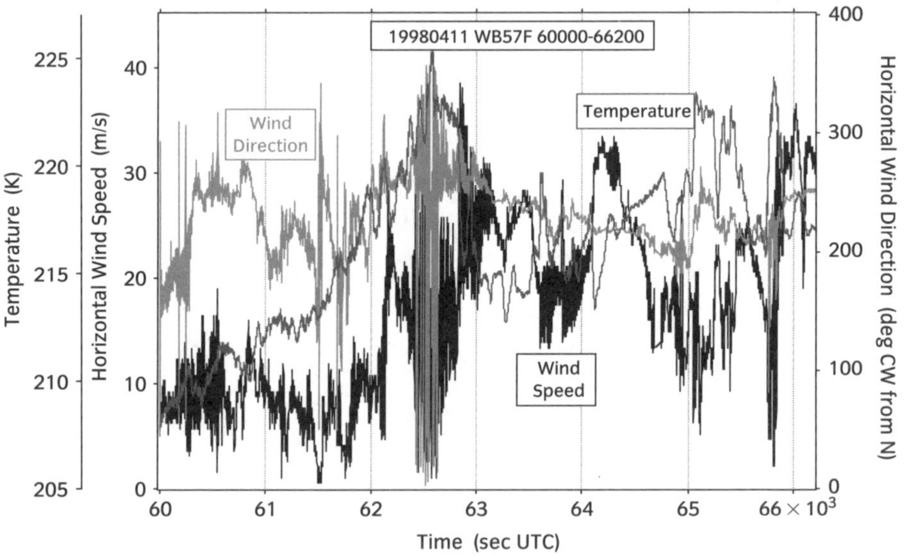

Figure 4.14 Temperature, horizontal wind speed, and wind direction traces for the northernmost section of the WB57F flight of 19980411, for the same isentropic 'curtain' shown in Figure 4.13. Note the high intermittency and turbulence in general and particularly near 62 500 and 65 900 s UTC, corresponding to mountain wave events 'D' and 'I' in Figure 22.

Table 4.2 Scaling properties computed using PTW temperature and INS wind speed observations during WAM and the MM5 temperature and INS wind speed interpolated to the WAM flight track at 9.4 km grid spacing. Results from the WAM observations have been computed using only the range of scales available in the interpolated MM5 data. H_1 indicates persistence; C_1 indicates intermittency; both range from 0 to 1. Note that if a process is scale-invariant and is non-stationary increments, then $\beta = 2H_2 + 1$. Multifractality or multiscaling is indicated by the non-constant values of $H(q)$; values of H_1, H_2, and H_6 shown are evidence of the multifractality of all four data sets.

Data set	H_1	H_2	H_6	β	$2H_2 + 1$	C_1
PTW Temperature	0.36	0.33	0.23	1.59	1.66	0.05
MM5 Temperature	0.64	0.58	0.50	2.00	2.16	0.09
INS Wind Speed	0.29	0.25	0.15	1.83	1.5	0.05
MM5 Wind Speed	0.58	0.53	0.45	2.05	2.06	0.09

altitude of the isentropes is real, not instrumental. The locations marked in Figure 4.13 by points D and I on the isentrope curtain are correlated with violent fluctuations in both wind speed and direction, coincident with near vertical isentropes over a considerable depth of the atmosphere. The temperature and wind traces did show scale invariance, see Table 4.2.

It was decided to compare the scaling behaviour of the observations with that in a simulation of this event by the MM5 model, the inner domain

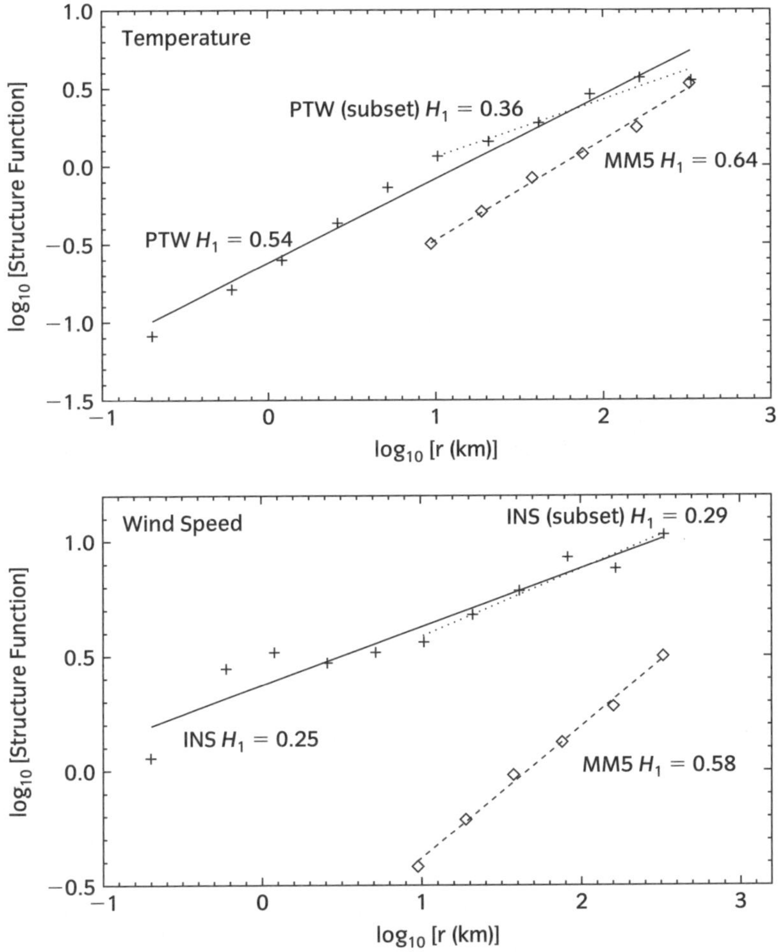

Figure 4.15 Log-log plots of interval distance versus structure function used in calculation of scaling exponents H_1. Top: WB57F observed temperature (plusses; solid line) and interpolated MM5 (diamonds; dashed line) numerical model temperatures along the aircraft flight track. Bottom: aircraft inertial navigation system wind speed (plusses; solid line) and interpolated MM5 (diamonds; dashed line) numerical model horizontal wind speeds. Least squares fits and H_1 are also shown. The dotted line indicates a fit to observational structure functions using only MM5 scales. Although the MM5 model does show scaling over its limited scaling range, the scaling exponent values do not agree with those from the observations.

(of three) of which was a square bounded by approximately (47°N, 111°W); (46°N, 103°W); (40°N, 105°W); (41°N, 113°W) having a horizontal resolution of 6.67 km and 41 vertical levels between the surface and 30 hPa. It was initialized with the ECMWF (European Centre for Medium-range Weather Forecasts) T106 analysis at 00 UTC on 11 April 1998 and was integrated for 18 hours. Two simple points are evident from Figure 4.15: the model data 'along the flight track' do scale for both wind speed and temperature, but with values considerably different from the aircraft data, for both H_1 and C_1. There were insufficient data to calculate α for either data set.

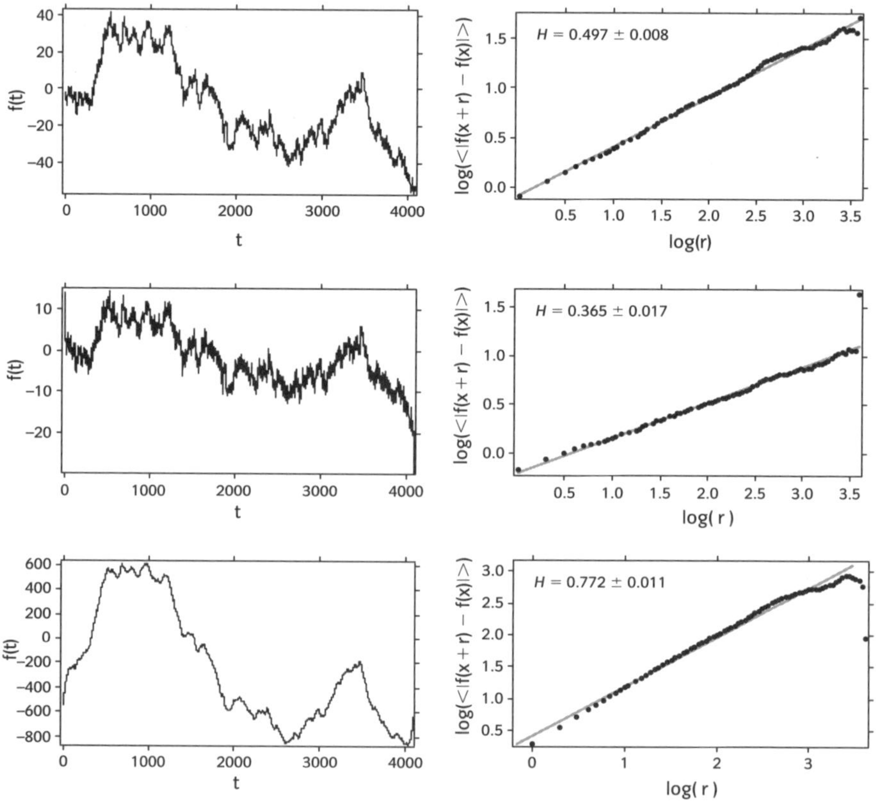

Figure 4.16 Examples of simulated signals with particular scaling behaviours, and their associated log-log plots. Upper panels, a synthetic signal generated by integrating Gaussian noise, which has $\beta = 0$; such an integration always increases β by 2, so yielding $H = (\beta/2) - 1 = 0.5$. Neighbouring points are uncorrelated. Middle panels, a synthetic signal generated by spectral filtering to produce a signal with $\beta < 2$. Here the neighbouring points tend to be anticorrelated, the trace is rougher and $H < 0.5$. Bottom panels, a synthetic signal generated by spectral filtering to produce a signal with $\beta > 2$; here the neighbouring points tend to be correlated and the trace has become smooth. The upper, middle, and lower cases are respectively referred to as random, antipersistent, and persistent. These terms should be reserved for Gaussian processes and avoided for Lévy and multifractal processes, where intermittency is significant.

The general simulation of the structures at D and I in Figure 4.13 did succeed in producing gravity waves, but the breaking extents were underestimated, particularly above 20 km, and the vertical extent was only ~3 km versus at least a scale height in the observations.

It may be that H_1 and C_1 from generalized scale invariance analysis of observations could be applied to numerical model simulations to improve the parametrizations of the smaller scales. In this context, a single scaling exponent, when used in a cloud microphysical model to simulate temperature variability made a significant difference to ice formation (Murphy 2003). Figure 4.16 gives an example of the simulation of a time series using $H_1 = 0.6$, while Figure 4.17 gives an example using typical values of H_1, C_1 and α. Figure 4.17, although more realistic than Figure 4.16, still does not quite capture the texture of the real aircraft data, looking too regular.

The 'crinkliness' of the isentropes in Figure 4.13 is not limited to flights near mountains; it is ubiquitous, at amplitudes a factor of two less than over land, over the southern oceans thousands of kilometres from any terrain (Gary 1989; Murphy and Gary 1995; Gary 2006). We further note that such isentrope curtains offer the possibility of multipoint correlation approaches in the fractal analysis (Shraiman and Siggia 2000); it has not yet been attempted.

Finally, we add a footnote to observational analysis by statistical multifractal techniques. The usual assumption is that the trajectory of an aircraft

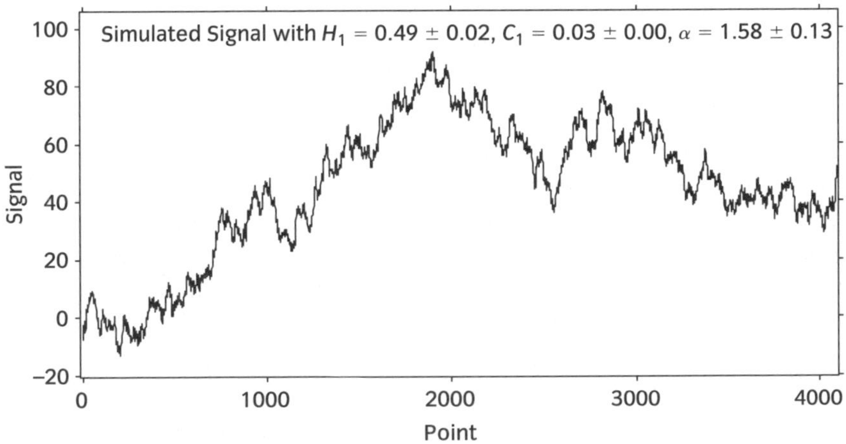

Figure 4.17 A simulated multifractal signal with values of H_1, C_1, and α typical of atmospheric data. Comparison of the signal with those in Figure 4.16 shows the simulated difference between Gaussian and Lévy processes. It is difficult to produce synthetic traces with great verisimilitude to real aircraft traces, such as those in Figures 4.1, 4.2, and 4.9, but Figure 4.17 is more realistic than Figure 4.16.

or dropsonde through the atmosphere is one-dimensional, making application of the intersection theorem simple: the codimension of the intersection of two fractal sets is the sum of the codimensions. In reality, as explained earlier, aircraft trajectories are fractal (Lovejoy et al. 2004); work has been done recently examining the trajectories of dropsondes, see Lovejoy et al. (2007a, b) and Hovde et al. (2007a). While the Hurst exponent, H_1, is robust in the sense of being tolerant of data gaps, this is not true of the intermittency, C_1, and the Lévy exponent, α. The data sets for C_1 and for α are more restricted, being limited to a subset of wind speed, temperature and ozone data for the polar vortices from ER-2 flights. Nevertheless, a first idea can be obtained of the parts of exponent space which are populated in the (H_1, C_1), (H_1, α), and (C_1, α) planes.

4.3 Polar lower stratosphere: H_1, C_1, and α

A generalized scale invariance analysis for all three exponents was possible for wind, temperature, and ozone for ER-2 great circle flight segments flown in the Arctic lower stratosphere during the POLARIS mission April–September 1997 and the SOLVE mission January–March 2000, with the ozone analysis also being possible for the ER-2 flights during the AAOE mission over west Antarctica in August–September 1987 and the AASE mission over the Arctic in January–February 1989 (Tuck et al. 2002, 2005). The algorithm for calculating the exponents shows that the three are closely related. For wind speed, $H_1 \approx 5/9$ implies that the dimensionality of atmospheric motion is 23/9, neither 2D ($H_1 \rightarrow 0$) nor 3D ($H_1 \rightarrow 1$). For a passive scalar, $H_1 \rightarrow 0$ implies complete decorrelation and $H_1 \rightarrow 1$ implies complete correlation; the observations of nitrous oxide agree with theory that $H_1 \approx 5/9$.

The intermittency C_1, a measure of the degree to which activity in a turbulent fluid is sporadic, is intimately associated with the long tails on probability distributions in which infrequent, high amplitude events make a significant contribution to the mean. When $C_1 \rightarrow 0$ the fluid approaches homogeneity, when $C_1 \rightarrow 1$ the energy concentrates in single structures. Whereas C_1 characterizes the sparseness of the mean of the field, α characterizes the distribution of the remaining values. For a Gaussian distribution, $\alpha \rightarrow 2$ ($\alpha' = 2$); for $1 < \alpha < 2$ ($2 < \alpha' < \infty$) the PDF is strongly asymmetric. It is particularly demanding on both quality and quantity of data to determine α accurately, as we have noted earlier.

Figure 4.18 shows results for wind speed, showing $0.37 < H_1(s) < 0.58$, $0.025 < C_1(s) < 0.042$ and $1.2 < \alpha(s) < 1.7$, with mean values 0.53, 0.036, and 1.43 respectively. These values are a considerable departure from Gaussianity in the inner Arctic vortex ($s < 30\,\mathrm{ms}^{-1}$); a wind speed

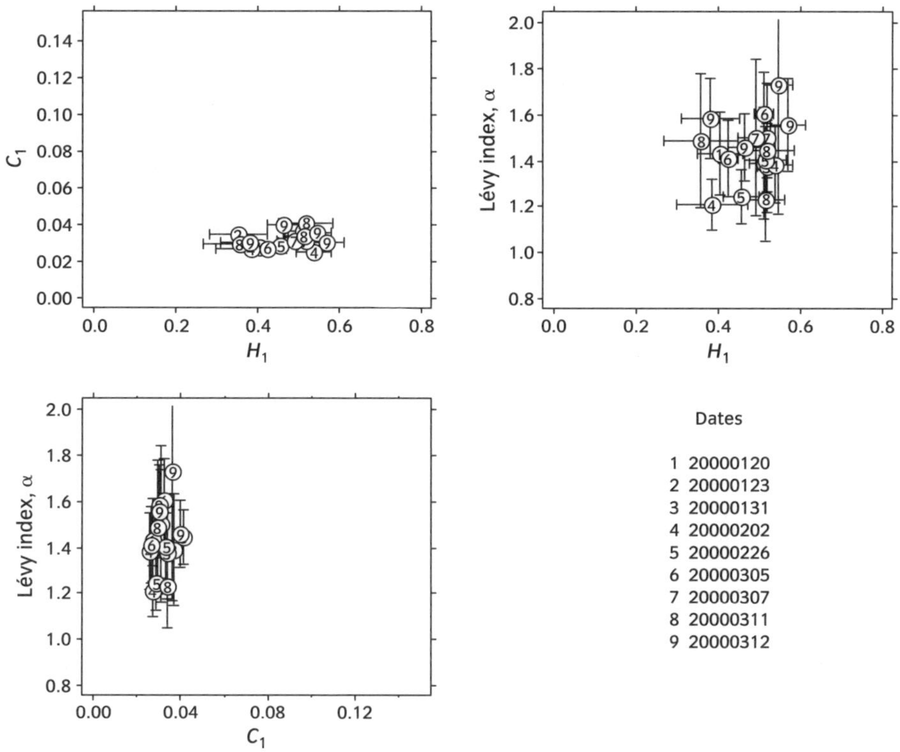

Figure 4.18 Generalized scale invariance analyses of wind speed during the SOLVE mission, inner vortex, plotted on (H_1, C_1), (H_1, α), and (C_1, α) exponent planes. The coding of the points proceeds from 1 in late January to 9 in mid-March 2000. The data were taken during 'horizontal' flight segments by the ER-2, largely inside the lower stratospheric polar vortex. The data are clearly multifractal and there is no significant time trend and no significant pairwise correlation among the three scaling exponents.

definition of the inner vortex's perimeter is used because it was directly measured by the aircraft instruments, and because it is consistent with the conventional definition of a jet stream. The results for temperature are shown in Figure 4.19. The result for $H_1(T)$ is close to both theory (5/9) and that for wind speed (0.53) in that it is 0.54. However, the value for $C_1(T)$ is 0.11 and the mean value for $\alpha(T)$, 1.55, has very large error bars. It turns out that the large $C_1(T)$ and large error bars on $\alpha(T)$ are caused by truncation error in the archived 1 Hz temperature data; we display the data here to demonstrate sensitivity to the data quality necessary to determine the exponents which largely describe the departures from Gaussian distributions in the long tails of the PDFs. The correct values, calculated by retaining full precision in averaging down from 10 Hz and 5 Hz to 1 Hz, may be seen in Table 4.3. Random instrumental noise was evident at 10 Hz but not at 5 Hz, which frequency gave results the same as those at 1 Hz obtained by averaging with full precision from 5 Hz. The data in

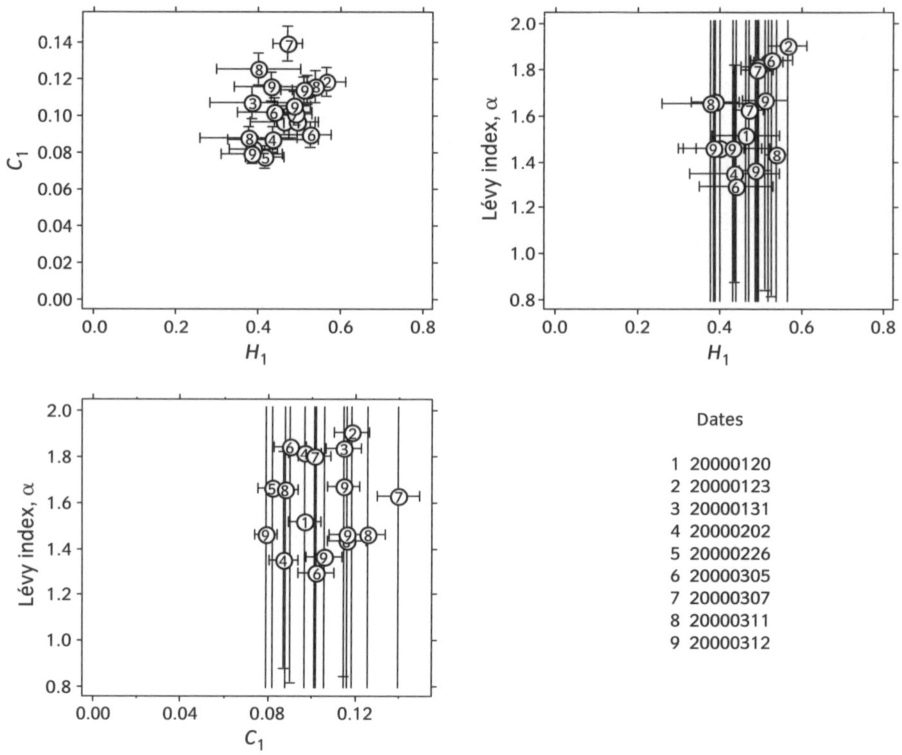

Figure 4.19 As Figure 4.18 but for temperature. These scaling exponents were calculated using the 1 Hz temperature data on the publicly available archive, which used single precision averaging down from the electronic sampling frequency of 300 Hz and which are shown to demonstrate the sensitivity of the intermittency C_1 and the Lévy exponent α to the precision of the data. The values of $C_1(T)$ are about a factor of three too large and the error bars on $\alpha(T)$ are exaggerated, compared to values obtained by averaging at double precision down from 10 Hz and 5 Hz data. The correct values may be seen in Table 4.3.

Table 4.3 show differences in the H exponent in the along-jet and across-jet directions, arguing for the greater effectiveness of speed shear over directional shear as an agent of decorrelation, that is in the promotion of mixing (Tuck et al. 2004). These H scaling exponents less than unity preclude the interpretation of the polar vortex as a 'containment vessel' for polar ozone loss, supporting the conclusions of Tuck (1989), Tuck et al. (1992), Rosenlof et al. (1997), Tuck et al. (1997), Tuck and Proffitt (1997), Tan et al. (1998), Plumb et al. (2000), Tuck et al. (2002) and Proffitt et al. (2003) that ozone loss and exchange with extra-vortex air were occurring at about the same rate, typically a few percent per day. Chemical loss of ozone proceeded at 2–4% per day, with concurrent mass loss from the vortex of 1–2% per day from August–October 1994 over Antarctica, with similar numbers over the Arctic during January–March 2000. There was detectable evolution over time of the scaling of wind speed and temperature

Table 4.3 Values of H_1, C_1, and α for wind speed, temperature, and ozone at 1 Hz with preservation of full precision. ER-2 data from SOLVE, January–March 2000.

Date (yyymmdd)	Start & stop times	$H_a(S) \pm$ c.i.	$C_1(S) \pm$ c.i.	$\alpha(S) \pm$ c.i.	$H_a(T) \pm$ c.i.	$C_1(T) \pm$ c.i.	$\alpha(T) \pm$ c.i.	$H_a(O_3) \pm$ c.i.	$C_1(O_3) \pm$ c.i.	$\alpha(O_3) \pm$ c.i.
20000111*	50118–54288	0.45 ± 0.05	0.042 ± 0.002	1.45 ± 0.33	0.40 ± 0.10	0.088 ± 0.006	1.34 ± 0.94	0.39 ± 0.06	0.070 ± 0.002	1.80 ± 0.15
20000111*	54288–59538	0.43 ± 0.05	0.060 ± 0.001	2.31 ± 0.74	0.39 ± 0.08	0.102 ± 0.004	1.48 ± 0.84	0.44 ± 0.05	0.052 ± 0.002	1.89 ± 0.15
20000114	46800–63931	0.52 ± 0.04	0.034 ± 0.001	1.72 ± 0.14	0.59 ± 0.05	0.067 ± 0.004	1.48 ± 0.56	0.44 ± 0.06	—	—
20000120	37553–47828	0.47 ± 0.06	0.026 ± 0.002	1.53 ± 0.21	0.48 ± 0.06	0.094 ± 0.007	1.78 ± 0.96	0.31 ± 0.04	0.023 ± 0.002	1.56 ± 0.15
20000123*	31017–38648	0.41 ± 0.05	0.027 ± 0.002	2.22 ± 1.39	0.57 ± 0.05	0.096 ± 0.007	1.84 ± 1.02	0.31 ± 0.06	0.023 ± 0.002	2.21 ± 0.92
20000127	45647–52267	0.43 ± 0.07	0.044 ± 0.002	2.38 ± 1.30	0.44 ± 0.07	0.085 ± 0.005	1.91 ± 1.97	0.41 ± 0.05	—	—
20000131	38199–42764	0.54 ± 0.05	0.037 ± 0.002	1.39 ± 0.24	0.54 ± 0.04	0.114 ± 0.008	1.83 ± 1.77	0.37 ± 0.07	0.027 ± 0.002	1.54 ± 0.15
20000202	42000–53500	0.54 ± 0.04	0.026 ± 0.002	1.37 ± 0.16	0.51 ± 0.05	0.094 ± 0.007	1.72 ± 1.85	0.41 ± 0.06	0.023 ± 0.002	1.39 ± 0.13
20000203	69353–72748	0.39 ± 0.04	0.038 ± 0.003	1.46 ± 0.22	0.31 ± 0.10	0.064 ± 0.005	1.11 ± 0.33	0.38 ± 0.08	0.037 ± 0.002	1.54 ± 0.25
20000226	31000–43000	0.57 ± 0.05	0.034 ± 0.002	1.38 ± 0.15	0.49 ± 0.05	0.076 ± 0.006	2.75 ± 2.00	0.39 ± 0.02	0.024 ± 0.002	1.53 ± 0.19
20000305	41747–52392	0.52 ± 0.03	0.032 ± 0.002	2.12 ± 1.23	0.55 ± 0.05	0.090 ± 0.006	2.30 ± 1.82	0.35 ± 0.06	0.026 ± 0.002	1.78 ± 0.21
20000307	37000–43000	0.46 ± 0.04	0.033 ± 0.002	1.43 ± 0.20	0.51 ± 0.05	0.104 ± 0.008	1.60 ± 1.00	0.26 ± 0.02	0.025 ± 0.002	1.33 ± 0.13
20000311	31524–39509	0.56 ± 0.06	0.035 ± 0.002	1.71 ± 0.63	0.60 ± 0.04	0.101 ± 0.008	1.59 ± 1.91	0.55 ± 0.04	0.042 ± 0.002	1.94 ± 0.23
20000311	42354–52389	0.62 ± 0.06	0.031 ± 0.002	1.55 ± 0.20	0.46 ± 0.05	0.093 ± 0.006	1.92 ± 1.71	0.55 ± 0.02	0.034 ± 0.002	1.82 ± 0.18
20000312	51934–58549	0.56 ± 0.04	0.030 ± 0.002	1.56 ± 0.18	0.44 ± 0.07	0.095 ± 0.007	1.66 ± 0.93	0.35 ± 0.03	0.032 ± 0.002	1.84 ± 0.21
20000316*	30048–55857	0.41 ± 0.04	0.027 ± 0.002	1.66 ± 0.30	0.45 ± 0.03	0.075 ± 0.005	1.48 ± 0.91	0.51 ± 0.02	0.034 ± 0.001	1.87 ± 0.22
20000318*	53934–69882	0.35 ± 0.06	—	—	0.41 ± 0.04	0.070 ± 0.005	1.60 ± 1.06	0.40 ± 0.04	0.041 ± 0.001	1.86 ± 0.22
Mean(across-jet)		0.51 ± 0.15	0.033 ± 0.011	1.63 ± 0.80	0.49 ± 0.18	0.090 ± 0 031	1.81 ± 1.26	0.40 ± 0.18	0.029 ± 0.014	1.63 ± 0.47
Mean(along-jet)		0.41 ± 0.10	0.039 ± 0.033	1.91 ± 1.26	0.44 ± 0.18	0.086 ± 0.029	1.55 ± 0.81	0.41 ± 0.17	0.044 ± 0.037	1.93 ± 0.53

* indicates an along-jet flight; all others are across-jet.

in the lower stratospheric polar vortex, as expected from the correlations of $H_1(s)$ and of $H_1(T)$ with measures of jet stream strength in Figure 4.9; the seasonal waxing and waning of the polar night jet stream and the vortex from autumn through winter to spring is reflected in the temporal behaviour of these exponents, see Figure 6 of Tuck et al. (2004). This result suggests that large-scale structures like jet streams are encompassed by generalized scale invariance, but at the expense of the universality of the numerical values of the scaling exponents. Pure two-dimensional turbulence should be independent of planetary rotation and can therefore be excluded in the atmosphere; the fact that $H_1(s)$ shows correlation with jet stream strength is consistent with the observed intermediate 23/9 dimensionality, since the Coriolis parameter appears in the thermal wind equation.

There is also temporal evolution of the ozone scaling, in both the Arctic and the Antarctic, as shown in Figures 4.20 and 4.21 respectively. The

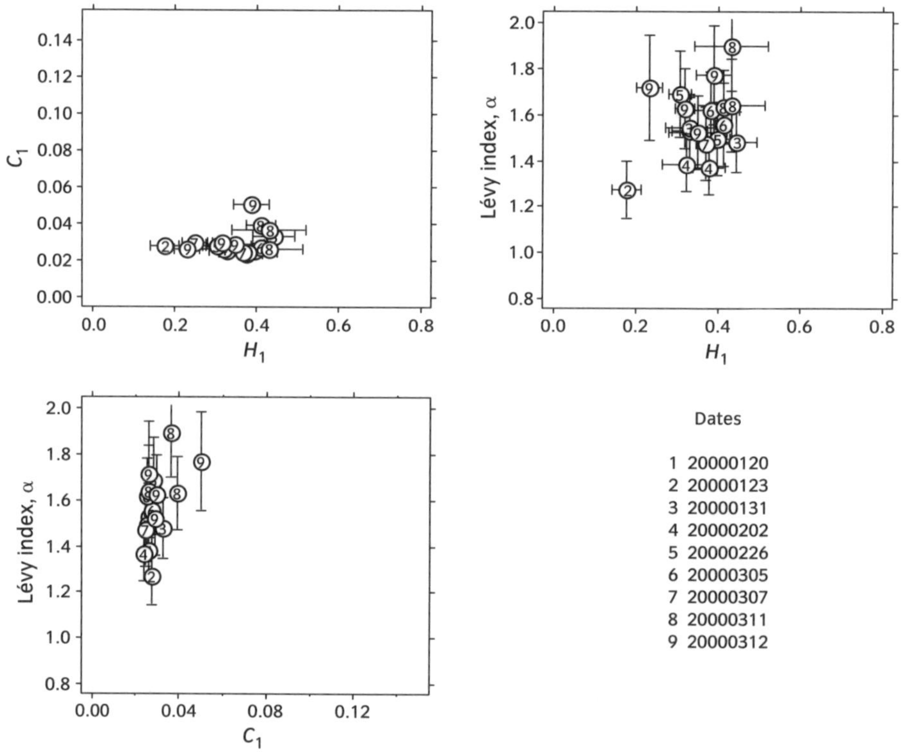

Figure 4.20 As Figure 4.18 but for ozone. During the 72-day period, there has been an increase in the Lévy exponent, α, from 1.25 in late January to 1.88 in mid-March which is significant. There are also smaller, less coherent increases in the intermittency, C_1, and in the conservation exponent, H_1. There is no clear guidance from theory about what is predicted or expected from chemistry in the turbulent atmosphere, but it is clear that there are detectable changes in the ozone scaling as the result of the polar photochemical sink processes. See text for discussion.

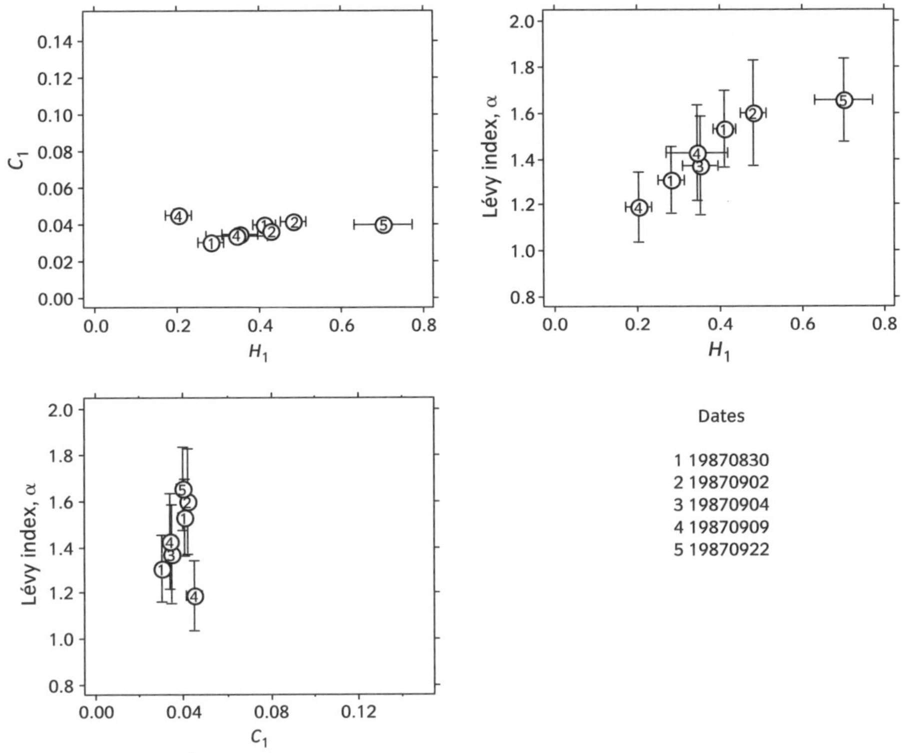

Figure 4.21 Ozone during the AAOE mission, ER-2 data taken in the Antarctic inner vortex near the Antarctic Peninsula, plotted on (H_1, C_1), (H_1, α), and (C_1, α) exponent planes. The numbering proceeds from 1 in late August to 9 in late September. There was a correlated increase in both the Lévy exponent, α, and the conservation exponent, H_1, during September while the intermittency, C_1, stayed steady. See text.

Arctic inner vortex ozone, as sampled by the ER-2, showed temporal evolution of its Lévy exponent between January and March 2000, increasing within the range $1.25 < \alpha < 1.88$. This has been interpreted as the introduction of air filaments from the outer vortex containing ozone mixing ratios that lead to a greater variety of ozone variations, as ozone loss proceeded from mid-winter to spring (Tuck et al. 2002). This will be further discussed in Section 6.2 in the light of the scaling behaviour of reactive chlorine and reactive nitrogen. In the Antarctic, as the 1987 ozone hole evolved from late August to late September, the scaling exponents $H_1(O_3)$ and $\alpha(O_3)$ both showed significant, and correlated, increases. The intermittency of ozone on the other hand, was stable in the range $0.03 < C_1(O_3) < 0.05$. There is no clear theoretical guidance on what scaling behaviour to expect from the interplay of photochemical kinetics and fluid dynamics; it is clearly not that of a passive scalar, and should be a useful observational test for numerical model simulations of the ozone hole. The generality of maximum entropy production and the fluctuation theorem (Dewar 2003; 2005a,b) suggests commonality in process, a notion consistent with the fundamental dependence of both chemical kinetics and turbulent vorticity upon molecular velocity distributions. Alder and Wainwright (1970) showed that fluid mechanical behaviour emerges in the form of 'ring currents' from molecular populations subjected to an anisotropy; vorticity has molecular causes, vorticity describes turbulence and turbulence affects the chemical kinetics in the atmosphere by virtue of the fluctuations it causes in the chemical concentrations involved in the rate equation via the law of mass action, as we shall see in Chapter 6.

4.4 Summary

We have seen that substantial volumes of in situ observations of wind and temperature exhibit generalized statistical multifractality while the canonical value of $H_1 = 5/9$ holds for the horizontal observations; this is not so in the vertical. An unexpected correlation for wind speed, s, emerges in both the horizontal and vertical for jet streams: $H_1(s)$ is correlated with measures of the jet strength, suggesting a connection with large-scale meteorology. This is also true for $H_1(T)$ in the horizontal, but not in the vertical. Extensive observations suggest that N_2O, a conserved tracer in the domain of observation, closely obeys the expected value of 5/9 for a passive scalar. On many flights and descents, ozone and water do not, suggesting the operation of sources and/or sinks. One case study showed that while MM5 high spatial resolution model data scaled, in the case of an aircraft flight in very turbulent air in the lee of the Rocky Mountains, the exponents were not in agreement with the observational scaling. All of these results have the potential to act as new and quite general observational tests of numerical models of the atmosphere, for both the dynamical and the chemical variables. Because the atmosphere does not yield Gaussian or Poisson distributions of variables along flight tracks, while such distributions are characteristic of most instrumental noise, there is also an objective test to determine the frequency at and above which the observations are significantly affected by instrumental noise, evident as a flattening of the slope of the variogram at the smallest scales.

5 Temperature Intermittency and Ozone Photodissociation

During the last two missions performed by the ER-2 in the Arctic lower stratosphere, POLARIS in the summer of 1997 and SOLVE during the winter of 1999–2000, an unexpected correlation emerged when the data were subjected to analysis by generalized scale invariance. It was between the intermittency of temperature, a number which can be determined for each segment of analysable flight from the temperature measurements, and the average over the flight segment of the photodissociation rate of ozone, which was calculable as a time series along the flight segment by taking the product of the 1 Hz measurements of the local ozone concentration and the 1 Hz measurements of the ozone photodissociation coefficient. In searching for a physical explanation of this correlation, it was realized that the common link between the quantities was that ozone photodissociation produces photofragments of atomic and molecular oxygen that recoil very fast, while temperature itself is the integral of the translational energy of all air molecules. The next step therefore was to ask if the intermittency of temperature was correlated with the average of the temperature itself over the flight segment: it was. One might think that because ozone is present at about 20 km altitude in mixing ratios of about $2 - 3 \times 10^{-6}$, the rapid quenching of the translational energies of the recoiling photofragments by molecular nitrogen and molecular oxygen would prevent any possible effects from showing up in the bulk, observed temperature. However, during the POLARIS mission, it was possible to fly the ER-2 near the terminator, the boundary between day and night, because at Arctic latitudes the planet was rotating slowly enough that it could fly legs in the same, stagnant air mass in both sunlight and darkness. These flights showed that the heating rate was significant, about 0.2 K per hour, and since heating in the stratosphere arises from the absorption of solar radiation by ozone, which leads to photodissociation, there is a prima facie case for considering non-local thermodynamic equilibrium effects from the recoiling fast photofragments. Two arguments may be deployed at this point, both from the theoretical literature; there are as yet no experiments on the translational speed distributions of atmospheric molecules. One is a calculation of the speeds of the atomic oxygen fragments after ozone absorbs a photon in the 200–300 nm wavelength region; they are moving very fast, up to 4000 ms^{-1}, compared to the 390 ms^{-1} of an average nitrogen molecule at 200 K (Baloïtcha and Balint-Kurti 2005). The second is the result that during molecular dynamics simulations, when a Maxwellian gas is subjected to an anisotropy—such as might be formed by the solar beam in our case—the fastest molecules respond by forming 'ring currents' on very short time and space scales (Alder and Wainwright 1970). These are fluid mechanical vortices, and are characterized by molecular speed distributions with over-populated fast tails. The ozone photofragments will not be recoiling into thermalized Maxwellian speed distributions but into the vorticity structures the fast molecules themselves have helped to create and sustain. If this hypothesized picture is correct, it has fundamental implications for the molecular interpretation of temperature in the atmosphere, even though the inertia of calibrated thermometers does result in a useable temperature record.

5.1 The Arctic lower stratosphere

The Arctic summer 1997 and winter 2000 missions saw the ER-2 equipped with an instrument designed to observe the spectrally resolved photodissociation rate of ozone, from measurements of the solar flux over the whole 4π solid angle and the ozone abundance between the sun and the ER-2 (McElroy 1995; Swartz et al. 1999). It turned out that the ozone photodissociation rate was correlated with the intermittency of temperature, an observation calling for an explanation.

We show an example of a time series along an ER-2 flight track in the Arctic summer anticyclone, and the calculation of $H_1(T)$ and $C_1(T)$ from it, in Figure 5.1. The flight was a long one, involving decreases of west longitude as well as latitude increases from Fairbanks, (65°N, 148°W). Notice that the temperature range during the outbound segment is relatively small, 4 K, but nevertheless the variability is real, multifractal and intermittent over more than four decades of horizontal scale. We take all such 'horizontal' flight segments from the POLARIS and SOLVE missions, some with temperature ranges up to about 20 K, and evaluate $C_1(T)$ in this way for each (NOTE: we use 5 Hz data, the shortest time interval free of instrumental noise, at full precision; anything less than full precision seriously affects the intermittency calculation, see Figure 4.19). For the same flight segments, we determine the average value of $J[O_3]$ for each, by taking the product of J and $[O_3]$ every second and computing the mean of the product over the flight segment. The resulting plot of $C_1(T)$ vs $J[O_3]$ is shown in Figure 5.2. The higher is the rate at which ozone is photodissociating, the more intermittent is the temperature. We can see from Equation 3.1 in

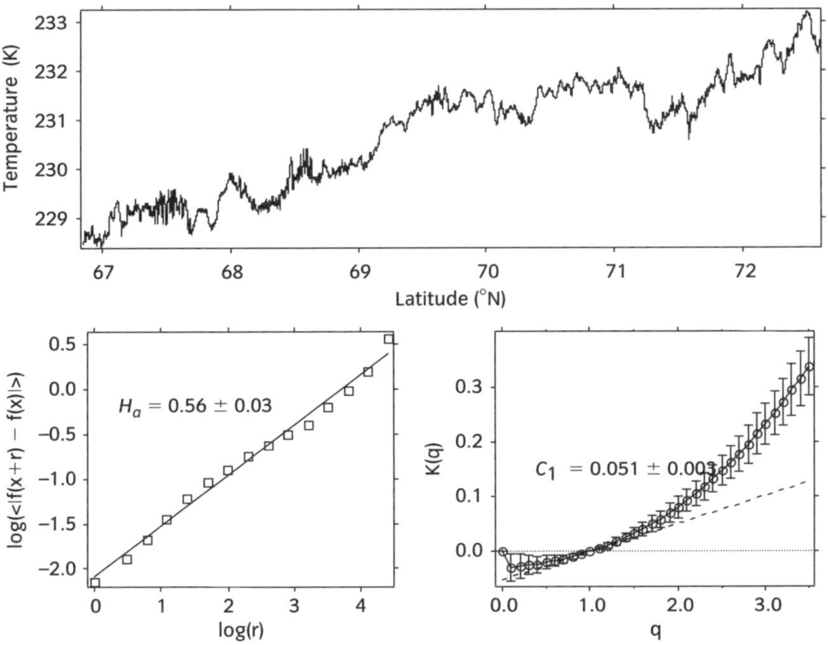

Figure 5.1 These are temperature data taken by the Meteorological Measuring System during an ER-2 flight into the Arctic summer anticyclone on 19970506 from Fairbanks (65°N, 148°W) towards a north-eastern direction; the flight segment was a great circle taking approximately 4700 seconds. MMS data were sampled electronically at 300 Hz. Averaging to 10 Hz showed effects of instrumental noise at small scales (flattening of the slope of the log-log plot when log(r)→0), not shown, while averaging to 5 Hz with full precision retained yielded stable intermittencies, C_1, free of random instrumental noise. Note $H = 0.56 = 5/9$ confirms Schertzer and Lovejoy Generalized Scale Invariance theory, because the aircraft samples in both the vertical and horizontal when under autopilot-controlled cruise climb.

Section 3.1 that in molecular terms temperature is

$$T = \frac{m}{3k_B}\left\langle (v - \langle v \rangle)^2 \right\rangle \tag{5.1}$$

and since ozone photodissociation is the primary source of translationally hot (fast moving) atoms, we might suspect that because intermittency is a reflection of a long tail in the temperature PDF (see Figure 4.1, Section 4.1), the intermittency $C_1(T)$ should also be correlated with T itself. This is so because transferring molecules from the most probable parts of their speed PDF to the fast tail will increase the average over the population, that is to say it will get hotter. This turns out to be observable via the effect of solar photons on temperature in the case shown in Figure 5.3. There are some Arctic summer flights which allowed a direct demonstration of the movement of population from the most probable regions of the temperature PDF to the warm tail. These were race track flights at 65°N in which

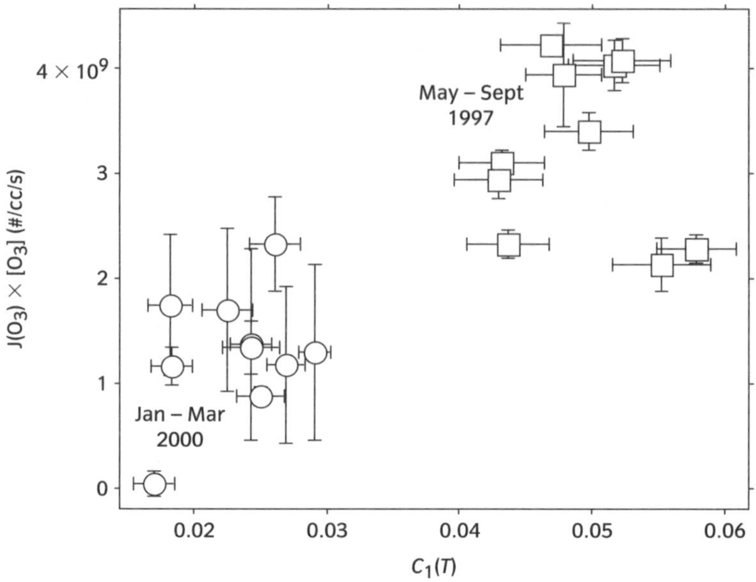

Figure 5.2 Ozone photodissociation rate (ordinate) and intermittency of temperature (abscissa) for the Arctic summer, 1997 (POLARIS), and the Arctic winter, 2000 (SOLVE). The data points are averages of $J[O_3]$ over the flight segment, with the vertical bars being standard deviations. $C_1(T)$ is evaluated for the flight segment, with the horizontal bar representing the standard error of the slope of the line fit. This correlation of the ozone photodissociation rate with the intermittency of temperature is discussed in the text.

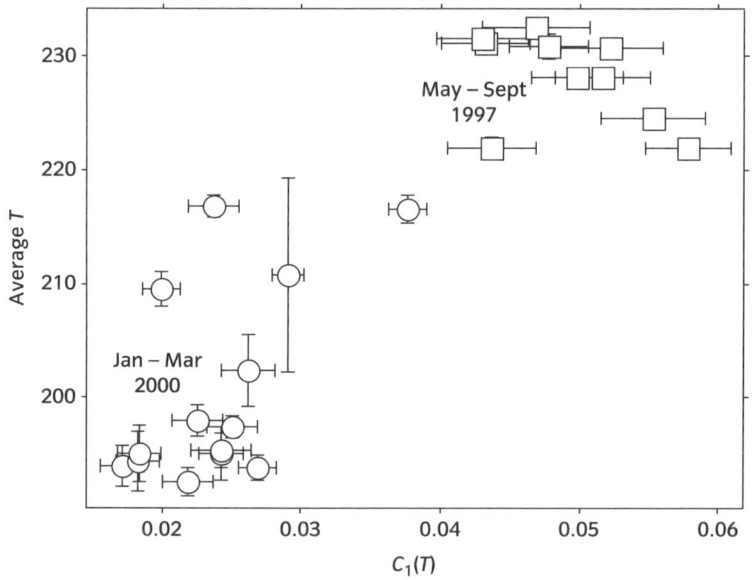

Figure 5.3 As Figure 5.2 but for average temperature rather than the ozone photodissociation rate. Since macroscopic temperature is proportional to the mean square molecular speed of the air molecules, this may be an indication that the speed distribution of the translationally 'hot' (fast-moving) photofragments from ozone is involved in the correlation. See text.

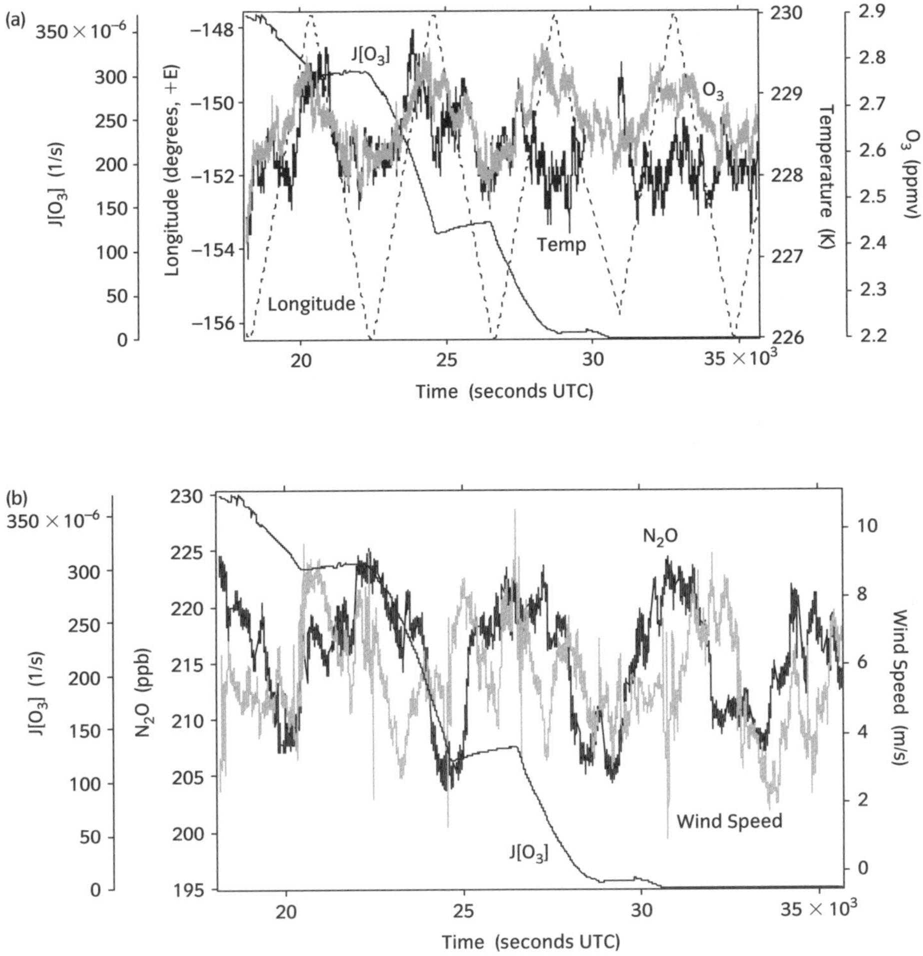

Figure 5.4 Flight data for 19970509 during POLARIS, race-track segments either side of the terminator at latitude 65°N in a slow-moving air mass. (a) longitude (dashed), $J[O_3]$ (thin black), ozone (black), and temperature (grey). (b) $J[O_3]$ (thin black), wind speed (grey), and nitrous oxide (black). Note that while ozone, nitrous oxide, and wind speed are approximately symmetrical about $J[O_3] \approx 0$, temperature is not; it is colder on the dark side of the terminator. Note that the wind speed, about $5\,\mathrm{ms}^{-1}$ from 040°, means that the air could cover no more than 3% of the race-track leg in the time the ER-2 completed it. The result confirms that absorption of solar radiation by ozone is large enough that the temperature itself has increased measurably, by 0.4 K in two hours at about 55 hPa in essentially the same air.

alternate segments were sunlit and dark, in 'the same air'. In Figure 5.4 it can be seen that while wind speed, ozone, and nitrous oxide are unchanged on either side of the terminator, this is not so for temperature. The PDF for temperature in the sunlit legs has lost population in the most probable regions and gained it in the warm tails in both May and September flights, see Figures 5.5 and 5.6. These results constitute a direct observation of

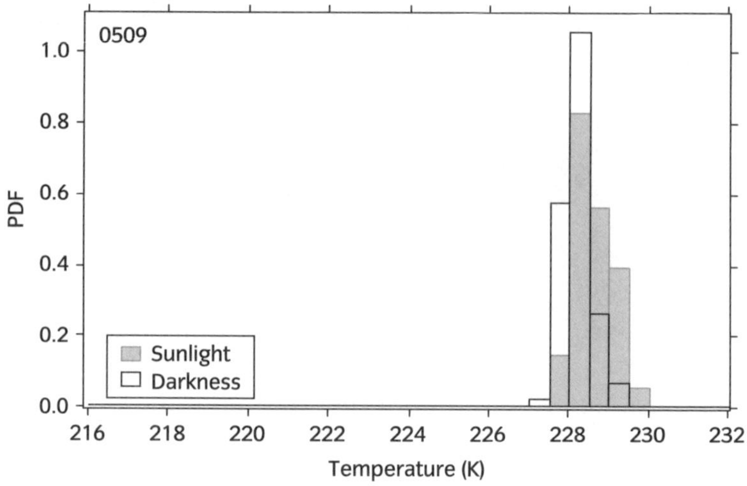

Figure 5.5 Probability distribution functions, 19970509, normalized to unity for the temperature data in Figure 5.2. Stippled, daylight; open, dark. Population has moved from the most probable values in the dark side to the warm tail in the sunlit side 'in the same air', constituting an observation of the heating rate.

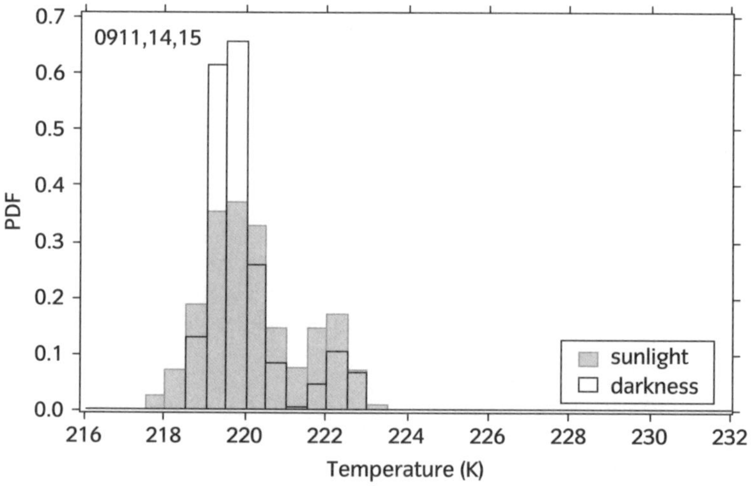

Figure 5.6 Probability distribution functions, 19970911-19970914-19970915, normalized to unity for temperature. The data were taken during race-track ER-2 segments in slow-moving air masses on either side of the terminator as in Figure 5.4 for May, but for three days in September. Stippled, daylight; open, dark. As in Figure 5.5, the sunlit data have gained population in the warm tail at the expense of the most probable values in the dark data. None of the race-track segments on the four days were long enough for a reliable evaluation of the intermittency of temperature.

the radiative heating, amounting to an average temperature rise of 0.4 K at 55 hPa in two hours as a result of $\approx 2 \times 10^9$ ozone photodissociations $cm^{-3}\,s^{-1}$ (Tuck et al. 2005). We note that other heating effects are smaller, a conclusion supported by the point having the lowest value of $C_1(T)$ in Figure 5.2—it is the only point virtually in the dark ($J[O_3] \rightarrow 0$). It permits an estimate of the contribution of infrared heating to the non-zero value of the temperature intermittency for this point, about 0.015. It should be noted that the correlation is not between $C_1(T)$ and either J or $[O_3]$ separately, but with their product, reinforcing the argument that it is the photodissociation *rate* which is significant. The other scaling exponents for temperature, $H_1(T)$ and $\alpha(T)$, showed no correlation with any of $C_1(T)$, T or $J[O_3]$, again supporting the argument and offering some refutation of the possibility of purely accidental correlation. An approximate calculation suggests that 2×10^9 ozone photodissociation events $cm^{-3}\,s^{-1}$ at 20,000 cm^{-1} photon energy input 1.4×10^{-9} kcal cm^{-3}. The energy needed to raise the temperature of air at 50 hPa by 0.4 K is 0.8×10^{-9} kcal cm^{-3}, so the energy requirements are compatible with the data shown in Figures 5.5 and 5.6. Perhaps the remainder went into producing non-equilibrium translational and rotational energy distributions in the air molecules.

What explanation can be offered for this correlation between a microscopic process and a macroscopic variable?

We recall that the energy of a 350 nm photon is approximately 29 000 cm^{-1}, while the bond dissociation energy of ozone is approximately 8 100 cm^{-1}, leaving over 20 000 cm^{-1} of available energy for the $O(^3P)$ and O_2 photofragments. Their translational energy could therefore be two orders of magnitude larger than k_BT. Molecular dynamics simulations have shown that hydrodynamic behaviour can be induced on time scales as short as 10^{-12} seconds and space scales as small as 10^{-8} metres (Alder and Wainwright 1970). The effect arises through the persistence of molecular velocity after collision; memory of initial velocity can persist for $\sim$100 collisions, causing molecules in the high velocity tail of the PDF to induce vortices, which occur in response to the faster particles inducing higher number densities ahead of them and lower number densities behind them. The general body of molecules will tend to equalize these number densities and in so doing create ‘ring currents’, or vortices. An important point is that temperature remains defined operationally to a good approximation because of the rapid equilibration of translational energy among the numerous molecules close in velocity to the most probable. The vortices and the overpopulation of high velocity molecules are mutually self-sustaining, a phenomenon which has many ramifications. An immediate implication is that it makes Maxwell–Boltzmann distributions of molecular velocities difficult to attain among air molecules, given that the atmosphere is never at rest, never equilibrated. The degree to which the rotational energy levels of the diatomic nitrogen and oxygen molecules that make up the great bulk of the atmosphere are equilibrated in 'ring currents' is an interesting and open question.

There are some eight orders of magnitude in horizontal distance between the largest scale covered by molecular dynamics simulations and the smallest (40 m) resolved by 5 Hz observations made on the ER-2. Nevertheless, we have observed correlation between a molecular scale process, ozone photodissociation, and a macroscopic fluid scaling exponent, the intermittency of temperature. These results from long flight segments are also consistent with the redistribution of temperature (accompanied by no such redistribution of ozone, nitrous oxide, and wind speed) on the sunlit side of the terminator relative to the dark side, seen on racetrack segments in slow-moving air masses. Since temperature remains approximately defined, overpopulated high molecular velocity tails in the PDF will translate to a similar redistribution in temperature, as observed; recall that $T \propto \overline{v^2}$, where $\overline{v^2}$ is the mean square Maxwell–Boltzmann velocity. Note that the transfer of translational energy to the atmosphere's main heat bath molecule, molecular nitrogen, from the translationally hot oxygen atoms and molecules from ozone photodissociation is implied. The mutually sustaining nature of the 'ring currents' and the overpopulation of high velocity molecules mean that normal energy transfer rates as measured in the laboratory cannot be applied. This is true even if the laboratory apparatus is free of 'ring currents', something which is far from guaranteed given that virtually all laboratory kinetics and spectroscopy experiments involve some form of anisotropy or directional flux, be it a gas flow or a laser beam. Such comments are also true of in situ atmospheric airborne instruments and are likely to become important considerations when successively smaller scales are addressed, when the instrumental length will no longer be small compared to the scale of the distance swept out by the aircraft during the frequency response time.

Recently, there have been calculations based on the laboratory data of Takahashi et al. (2002) showing non-equilibrium abundances of $O(^1D)$ atoms in the stratosphere, ranging from a factor of 2 at 50 km to about 20% at 20 km. Kharchenko and Dalgarno (2004) solved the time-dependent Boltzmann equation for $O(^1D)$ atoms in the stratosphere and mesosphere to obtain 3–5% of nonthermal atoms, corresponding to non Maxwellian speed distributions for these atoms (not the air itself) which corresponded to effective absolute temperatures 14% and 33% higher than those of the ambient air at 25 and 50 km respectively. It should be noted that neither of these calculations embodies the generation of ring currents (vorticity) seen by Alder and Wainwright (1970) and hence do not include the non-linear amplification effects discussed here and elsewhere in this book. These effects include the recoil of translationally hot photofragments into vorticity structures and the enhanced ozone photodissociation rates in the regions of higher density of these ring currents, both of which are non-linearly amplifying and which, we show, affect the whole population of air molecules and so the observed intermittency of temperature.

It should not be thought that ozone photodissociation is the sole source of temperature intermittency, although it does appear to be dominant. In Figure 5.2, the flight segment with $J[O_3] \rightarrow 0$ does not have zero intermittency, implying that the intermittency does not vanish in the dark. Other sources of anisotropy in the lower stratosphere operative in the absence of sunlight are infrared radiation from below, larger scale flow, gravitation, planetary rotation, and interaction of the atmosphere with the surface. There is also another mechanism which is non-linear and which will be effective on the scale of the 'ring currents'. These vortices are generated by the faster molecules of the PDF; they induce higher number density ahead of them and leave lower number density behind them. The 'ring current' will have not only faster ozone (and air) molecules in it, they will have a higher ozone photodissociation rate ahead and a lower one behind, a further positive feedback via the production of the very fast moving photofragments. Such positive feedbacks are important in generating long-tailed PDFs and the associated long-range correlations.

If we are correct about the link between ozone photodissociation and intermittency of temperature, it implies that if there is a cascade of energy it is upscale, at least at smaller scales, because molecules and photons represent the smallest length scales operative in the atmospheric energy budget. We have showed elsewhere from scaling analyses of total water and ozone that energy-conserving cascades are unlikely in the atmosphere (Tuck et al. 2003b), because the ubiquitous scale invariant, turbulent structure in the wind field imposes itself on their number densities and hence upon their absorbances and emittances. Energy input must therefore be on all scales. Physically, it seems that the mutually sustaining interaction between high velocity molecules and the vortices they induce must be of fundamental importance to atmospheric turbulence. It is interesting that the importance of the turbulent transfer of heat (molecular velocity) can be seen to be a potentially central process by analysis at scales which are small in atmospheric terms. Remarkably, Eady (1950) concluded that the turbulent transfer of heat was fundamental to the general circulation of the atmosphere, by arguments based on large-scale hydrodynamics; cellular circulations, jet streams etc. were held to be secondary phenomena. Conceptually, it may be that turbulence should be viewed as the emergence of larger scale order from the more nearly random molecular motion at smaller scales, rather than as the production of random motion at smaller scales from larger scale hydrodynamic instability. Atmospheric turbulence has molecular roots.

The above discussion is interesting when seen in the context of Eady (1951) and Eady and Sawyer (1951); it may provide an actual turbulence-producing mechanism, although fundamentally acting from small scales up rather than from large scales down.

We may also note that the asymmetries in the summer and winter temperature distributions suggest that in the Arctic, lower stratosphere heating in summer and radiative cooling to space in winter produce most probable temperatures which are respectively warm and cold relative to the annual mean for the region, with long tails of the opposite sign. The winter time distribution is also shared by the cold tropical tropopause; it is not clear exactly what mechanisms cause the long warm tails in these PDFs, or the long cool one in the Arctic summer, but they must be linked to the turbulent transport of heat. It is not known over what volume of atmosphere one would have to average for what period of time to obtain a Gaussian distribution of temperature, if one ever would—because the atmosphere is not at equilibrium. Indeed one could argue that because it is a system far from equilibrium a skewed distribution, relative to a Gaussian, should be expected. The experimental testing of this idea would involve the acquisition of high quality in situ data on a hitherto unprecedented scale. However, while such an effort was not technically feasible until very recently, it is now. Long-range high altitude unpiloted aircraft, carrying in situ instruments and dropping modern GPS sondes, are capable of executing such a programme (MacDonald 2005). The result would be a much more detailed understanding, underpinned by observation, of the energy state of the atmosphere and the processes which maintain it.

5.2 What is atmospheric temperature?

Temperature is defined macroscopically by Equation 3.16 and microscopically by Equation 4.19. In both cases, it refers to an equilibrated state. Gallavotti (1999) pointed out that in the macroscopic case, there is no formal definition of entropy far from equilibrium, a statement which implies difficulty in defining temperature via Eq. 3.16! On the other hand, Dewar (2003, 2005a, b) has derived a scale-free fluctuation-dissipation theorem via Jaynes' information theory formulation of non-equilibrium statistical mechanics, applying the maximization of entropy production. One result of this theorem is that temperature remains defined far from equilibrium. This result is of course in accord with macroscopic experience in the atmosphere. The microscopic perspective is found in Alder's (2002) molecular dynamics approach. An initially equilibrated population of molecules, with its velocity distribution described by a Gaussian, produces an overpopulated high velocity tail when subject to an external, disequilibrating perturbation. This is expressible in terms of power law decay of the molecular velocity autocorrelation function, Eq. 3.3. However, it is extremely difficult to obtain an analytical expression for the resulting long-tailed PDF of molecular velocities; numerical simulations show that temperature does remain well defined because of rapid equilibration of velocity among the many molecules with similar velocities which populate the high-probability regions of the PDF. The high-speed molecules, however, preferentially interact with each other to produce vortices, as described in Section 3.1. The vortices, once produced, maintain a mutually self-sustaining interaction with the overpopulated, high-speed tail. Thus temperature will still be an integral over the kinetic energy in the molecular velocities, but it will not be over a Gaussian; it will be whatever average the inertia of the 'thermometer' produces. The role of the two degrees of rotational freedom possessed by each of N_2 and O_2 poses some interesting questions in our context, upon which little or no observational evidence is available. It is not clear, in the non-equilibrium state of the air that we have posited, how readily these rotational energies will convert to translational energy.

The macroscopic relationship between vorticity and the thermodynamic state of a gas flow was derived and investigated by Truesdell (1952), and is sometimes given in meteorological texts as the Beltrami equation (e.g. Dutton, 1986). There appears to be no microscopic equivalent, but if molecular populations behave as fluids it ought to be possible to express vorticity in terms of the molecular velocity fields via Equations (3.9 to 3.13) after substituting molecular velocities v for fluid velocities u. In this case,

$$\langle \omega \cdot \omega' \rangle = -\nabla \langle v \cdot v' \rangle \tag{5.2}$$

which yields twice the enstrophy after taking the curl of the molecular velocity field. The vorticity of 'air' in a molecular dynamics simulation would then arise from

$$\omega = \nabla \times v\,(p, q) \tag{5.3}$$

where p is molecular momentum and q is molecular position. q is necessary because the intermolecular force field for real molecules depends upon separation and angle, and because v in the atmosphere will depend upon position as a result of anisotropies arising from gravity, planetary rotation and the solar beam. In such a simulation, the expression for the nth moment of the molecular speed (Landau & Lifshitz 1980)

$$\overline{v^n} = \frac{2}{\sqrt{\pi}} \left(\frac{2kT}{m} \right) \Gamma \left(\frac{n+3}{2} \right) \tag{5.4}$$

would allow calculation of the structure functions used in generalized scale invariance. In practice it would be necessary to investigate the averaging of the discrete molecular velocities before derivatives could be taken; it should be the minimum possible, to avoid damage to any emerging scaling properties.

Section 4.2 shows that the 'horizontal' scaling from aircraft and the 'vertical' scaling from dropsondes of temperature differ substantially, far more so than is the case for wind speed and humidity, as was seen in Figures 4.3–4.8. We argued that this was a sign of the prevailing influence of gravity, related to the vertical temperature structure of the atmosphere by the hydrostatic equation. We can actually deduce observationally the vertical scale below which this is no longer true, for example when the effects of the turbulent structure on these smaller scales is sufficient for the dry adiabatic lapse rate, 9.8 K/km, to be exceeded. We note that if the temperature in an air column decreases with height at a faster rate, it becomes convectively

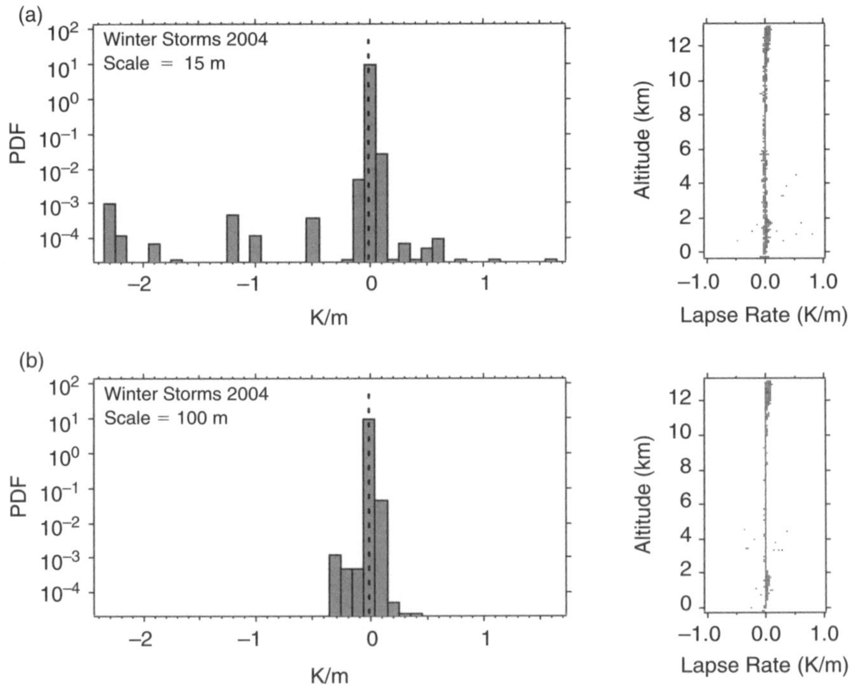

Figure 5.7 The PDFs for the value of the change of temperature with altitude (lapse rate; the dry adiabatic value of 9.8 K km^{-1} is marked by the vertical dashed line), accompanied by the vertical profile of these lapse rates. Winter Storms 2004 data from GPS dropsondes ejected from the NOAA G4 aircraft. The data were evaluated for (a), a vertical interval of 15 metres and (b), a vertical interval of 100 metres. Note that about 1% of the cases exceed the dry adiabatic lapse rate for the 15 metre interval and that none do for the 100 metre interval. Note that at 15 metre intervals, the tropopause is not evident, relative to the troposphere, whereas it is at 100 metre intervals.

unstable, with the too-warm air below changing places in the gravitational field with the too-cool air above. It can be seen from Figure 5.7 that at vertical scales below about 30 m the frequency of occurrence is no different in the troposphere and the stratosphere; the tropopause altitude is not evident. For vertical scales of 50 m and larger, the position of the tropopause is evident, in the form of consistently more stable lapse rates. The 50% exceedance of the dry adiabatic lapse rate occurs at a vertical scale of 93 cm, obtained by logarithmic extrapolation. This suggests that at these and smaller vertical scales, there may be some systematic effect where the small scale turbulence causes a steeper fall of temperature with height than purely random. Problems associated with fluid flow around the temperature sensor could also be responsible.

The PDFs of velocity and speed for Maxwellian molecules are shown in Figure 5.8; for an equilibrated gas the distribution of velocity must be symmetrical. The usual assumption in meteorology is that a Maxwell–Boltzmann speed distribution obtains at centimetric to millimetric scales and smaller, and is simply advected with the larger scale flow. This cannot be true in light of the molecular induction of vortices at scales of 10^{-8} m in times of 10^{-12} s, particularly when combined with the fact that observed wind speeds in the subtropical and polar night jet streams have reached 1/3 the most probable molecular velocities. How the equations which are integrated forward in time in large computers to forecast weather and climate could be modified to accommodate this reality remains to be seen, but it would seem that not only would an observationally based scheme have to be statistical and multifractal, that is to say scale invariant, it would have to incorporate molecular behaviour in a way that was realistic, at least in a stochastic sense, to achieve physically realistic prognosis of temperature.

The schematic coupling of molecular scale processes with weather and climate is shown in Figure 5.9. The molecular velocity is central; heat flux is primary to atmospheric motion. It is not clear if the microscopic and macroscopic definitions of temperature are consistent in the atmosphere, given that it is far from equilibrium and that the intermittency of measured, macroscopic temperature is correlated with the microscopic process of ozone photodissociation, at least in the lower stratosphere. It is also not clear for numerical models of the atmosphere. What is clear, however, is that the molecular interpretation of temperature will be different in an atmosphere with significantly altered number densities of the radiatively active gases, principally water vapour, carbon dioxide, and ozone. This will affect turbulence via molecular scale generation of vorticity, as we have seen, but increased collisional relative velocities, particularly among the faster moving molecules, will also impact radiative transfer via effects on spectral line shapes and will impact composition via effects on the rates of chemical reactions. A recent estimate from theoretical chemistry puts the translational velocity of the $O(^1D)$ atom from ozone photodissociation in

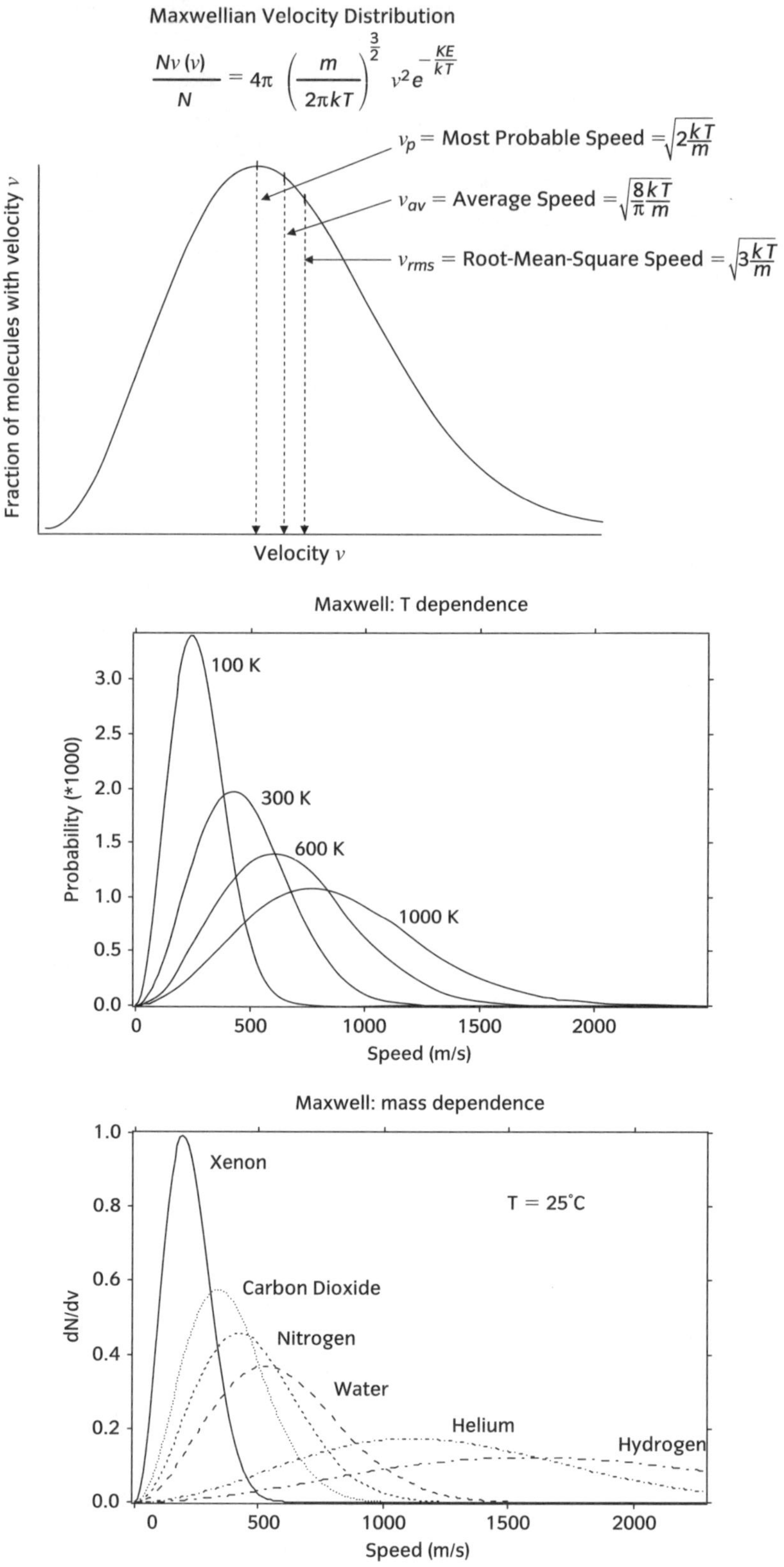

Figure 5.8 *Continued*

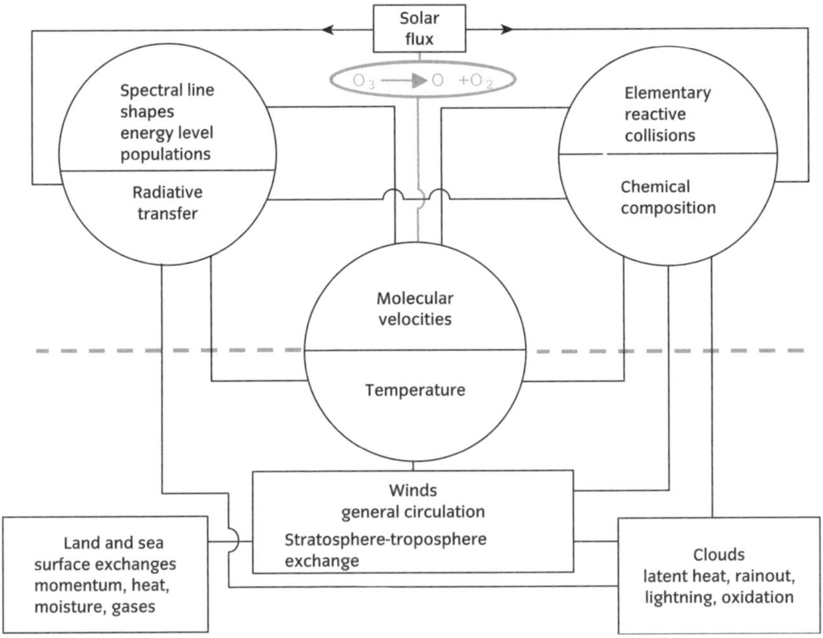

Figure 5.9 The top half of the diagram portrays microscopic (molecular) processes, while the bottom half is macroscopic (fluid mechanical and meteorological). For temperature and molecular velocity are central; while their relationship is known via the Maxwell-Boltzmann distribution for the statistical thermodynamics of an ideal gas at equilibrium, it is not known the non-equilibrium statistical mechanics needed in the case of the atmosphere.

the Hartley band at a most probable value of 2548 m s^{-1}, with a tail out to 4000 m s^{-1} (Baloïtcha and Balint-Kurti, 2005), see Figure 5.10. The more numerous ground state O(^{3}P) atoms—the flux is two orders of magnitude greater—could be moving faster; despite the lower photon energy, they potentially have nearly 16 000 cm^{-1} of extra energy available.

So, even though there is a reliable instrumental record of global surface temperature from the mid-nineteenth century to the present, its interpretation in molecular terms of what it means for turbulence, radiation, and chemistry will be different now than it was 130 years ago, particularly given the probable substantial increases in free tropospheric ozone since then (Volz and Kley 1988; Marenco et al. 1994; Harris et al. 1997), as may

Figure 5.8 Maxwellian speed distributions. The most probable speed of Maxwellian molecules with $m = 28$ (equivalent to N_2) at 200 K is 390 ms^{-1}. Wind speed in the subtropical jet stream and in the Antarctic stratospheric polar night jet can reach at least 130 ms^{-1}. (a), the speed distribution for $m = 28$ at 200 K. (b), the speed distributions for $m = 28$ at different temperatures. Note that the speed of atomic oxygen photofragments from ozone photodissociation can exceed 3500 ms^{-1}. (c), the mass dependence of Maxwellian speeds for masses of 1 (hydrogen atoms), 4 (helium), 18 (water), 28 (molecular nitrogen), 44 (carbon dioxide), and 131 (xenon).

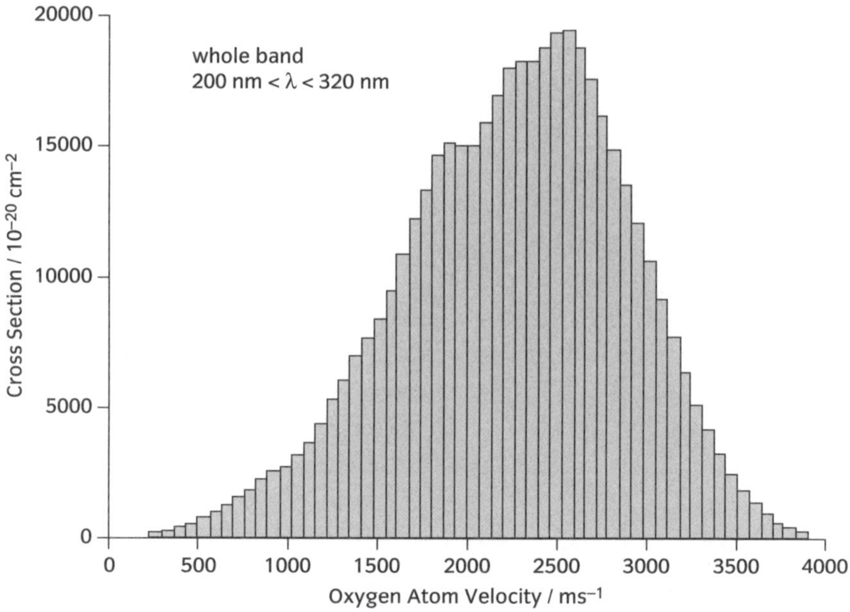

Figure 5.10 Velocity distribution of electronically excited $O(^1D)$ atoms produced when ozone photodissociates in the UV Hartley band, centred at 255 nm (Baloïtcha and Balint-Kurti, 2005). Ground state $O(^3P)$ atoms are also produced by solar photon absorption in the Huggins, Chappuis, and Wulf bands at longer wavelengths across the visible and into the near infrared; although the photon energy is less than in the Hartley band, there is nearly 2 eV of extra energy for translational excitation of the photofragments from ozone because the oxygen atom is produced in the ground state.

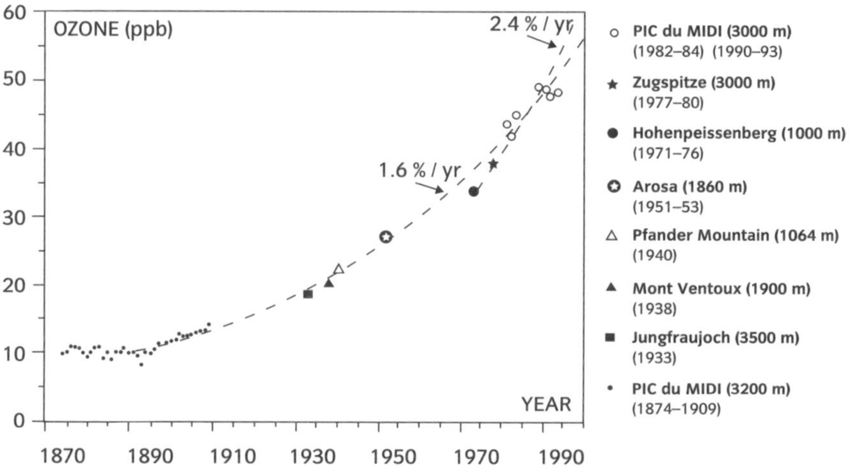

Figure 5.11 European time series of tropospheric ozone mixing ratio (Marenco et al. 1994). The sites are montane; the elevation is sufficient to ensure that free tropospheric air, above the boundary layer, was being observed. The increase means that there were many more hot photofragments from ozone at the end of the twentieth century than at the end of the nineteenth.

be seen in Figure 5.11. Overall, it seems likely that the sign of the effect would be to amplify the atmosphere's response to anthropogenic increases in so-called greenhouse gases, because more air molecules will be moving faster throughout the tropospheric column for a given solar irradiance at the top of the atmosphere, but this is a supposition concerning a non-linear system; some soundly based computations are required. The general cold bias in climate models (IPCC 2001) is noteworthy in this context.

Reading

Rapaport, D. C. (2004), *The Art of Molecular Dynamics Simulation*, 2nd edition, Cambridge University Press. Chapter 15 lays out some of the general potential difficulties likely to be encountered in a molecular dynamics simulation of ozone photodissociation in 'air'.

6 Radiative and Chemical Kinetic Implications

The laws governing the dynamical behaviour of atoms and molecules are quantum mechanical, and specify that their internal energy states are discrete, with only definite photon energies inducing transitions between them, subject to selection rules. These energy levels appear as spectra in different regions of the electromagnetic spectrum: pure rotational lines in the microwave or far infrared, 'rovibrational' (rotation + vibration) lines in the middle and near infrared, while electronic transitions, sometimes with associated rotational and vibrational structure ('rovibronic') occur from the near infrared through the visible to the ultraviolet. An important feature of these spectra in the atmosphere is that they do not appear as single sharp lines, but are collisionally broadened about the central energy into 'line shapes' which frequently overlap with other transitions, both from the same molecule and from others. One of the primary dynamical quantities involved in the processes broadening these line shapes is the relative velocity of the molecules with which the photon absorbing and emitting molecules are colliding. These are primarily N_2 and O_2 in the atmosphere; if they have an overpopulation of fast moving molecules relative to a Maxwell–Boltzmann distribution, as we have suggested, the line shapes will be affected. Molecules such as carbon dioxide, water vapour, and ozone are all active in the infrared via rovibrational transitions, with water vapour being light enough and so having sufficiently rapid rotation that it has rotational bands appearing in the far infrared rather than the microwave. Nitrous oxide, N_2O, and methane, CH_4, are also active, but make smaller contributions because of their lower abundances. Molecular nitrogen and molecular oxygen, because they are homonuclear diatomic molecules, do not absorb or emit via electric dipole allowed transitions in the atmospherically important regions of the electromagnetic spectrum. Molecular oxygen, having a triplet ground state, does have weak forbidden and magnetic dipole transitions which, however, play only a very small role in the radiative balance. It should be noted that the translational energy of molecules in a large system like the atmosphere is effectively continuous rather than quantized.

Atmospheric chemical reactions are also fundamentally affected by the relative velocities of reactant molecules as they enter a collision. Many reactions accelerate upon heating of the reactants. An increase in temperature increases the fraction of molecules moving fast enough to overcome the barrier ('activation energy') existing to the collisional transformation of a pair of reactant molecules to the product molecules. Some atmospheric reactions involving the recombination of atoms and free radicals—reactive molecular fragments with an unpaired electron—show a decrease in the rate of reaction at higher temperatures. This happens because the reaction can only occur if the briefly formed (of order 10^{-13} seconds) recombination product can dispose of the excess translational energy of the recombining reactants before flying apart. It does this by collision with any other molecule, and the lower the temperature, the easier it is because the less is the energy involved. Thus any change from a Maxwellian distribution of molecular speeds to a PDF with an overpopulation of translationally hot molecules could have a systematic effect on atmospheric chemistry, by accelerating reactions with activation energies and decelerating those with negative temperature dependences, largely recombinations of atoms and free radicals. This change would be conceptual and in our calculations, not in the real atmosphere of course. The isotopic composition of ozone in the stratosphere, which is not at quantum statistical equilibrium (Mauersberger 1981; Mauersberger et al. 1999, 2005; Gao and Marcus 2001), is an especially likely candidate to be affected by the mechanisms that have just been discussed, via the overpopulation of fast atoms and molecules involved in its three-body formation.

Molecular velocities when in an anisotropic environment such as the atmosphere give rise to turbulent vortices, as we have seen in Section 3.1. Because the vortices and the concomitant over-population of the high-speed tail of the molecular speed PDF are mutually self-sustaining, the effects will not be limited to those outlined above on radiative transfer and chemical kinetics. The turbulent structure of the air will be affected from the smallest scales up, so the coupling of dynamics, chemistry, and radiation is at a fundamental level, via the velocity distributions of the molecules comprising the atmosphere.

6.1 Radiative transfer implications

The shapes of the rovibrational spectral lines of water vapour, ozone, and carbon dioxide, across the infrared region of the electromagnetic spectrum, are fundamentally determined by the velocities of the air molecules with which these radiatively active molecules collide and the frequency at which they do so. At low pressures, say above the lower stratosphere, these lines have Doppler shapes—the broadening caused by the spread of molecular velocities. At higher pressures, the lines are principally broadened, from the single central energy representing the energy jump between the quantum levels concerned, by the effects of collisions. This is a very complicated quantum mechanical problem; the commonly used approximations are the Lorentzian and Voigt profiles, but deviations are observed in atmospheric spectra. For further reading, see Breene (1981) and Goody and Yung (1989). Thus whether we are considering a Doppler-broadened region such as the upper stratosphere, a Lorentzian region such as the troposphere, or a convolution of the two shapes in the form of a Voigt profile, the velocity distribution of the air molecules, not just the velocities of the absorbers and emitters, come into play. The harder the collision between an absorber and emitter molecule and a bath gas air molecule, the greater will be the perturbation to the quantum energy level separation involved in any spectral line. This will make the effects of an overpopulation of fast air molecules particularly effective in the far wings of the lines. These sorts of spectral region, where wings of lines overlap and absorb more weakly than near the line centres, are particularly effective at amplifying the effect of greenhouse gases, because they are less likely to be self-absorbed than the line centres. Recently, there has been experimental evidence that the relative velocities of the collidant gas molecules affect the line shape of water vapour lines in the ν_2 (asymmetric stretch) band (Wagner et al. 2005). The air broadening coefficients for lines with different rotational quantum numbers within the band showed different temperature dependences; although these data are not directly applicable to our problem of overpopulations of fast air molecules in the non-equilibrium, vorticity-laden environment of the atmosphere, they are a clear experimental indication of the reality of molecular speed effects upon the line shapes of the most effective 'greenhouse' gas. The water vapour continuum underlying the assignable lines in the monomer spectrum is probably caused by a combination of collisional broadening of these lines together with absorption by the dimer and higher clusters (Vaida et al. 2001). Each of these effects will respond differently to non-Maxwellian speed distributions in air molecules; neither is likely to be described with predictive power by fitting polynomials to differences between observed spectra and an hypothetical spectrum of ideal Lorentzian line shapes. Our purpose here is to point out the possible ramifications for calculation of atmospheric radiative transfer if the molecular speed distribution in the atmosphere depends upon the ozone photodissociation rate; we have observed that temperature intermittency is correlated with it.

We can illustrate the reason for concern with two diagrams, from very different parts of the literature. Figure 5.10 shows the velocity distribution of the electronically excited $O(^1D)$ atoms produced when ozone photodissociates in the UV Hartley band, centred at 255 nm (Baloïtcha and Balint-Kurti 2005). The speeds range up to 4000 ms^{-1}, a factor of over 10 greater than the most probable velocity of N_2 molecules at 200 K. When these excited photofragment atoms recoil, they do so into the complex, pre-existing field of vortices on all scales that characterize the atmosphere, whose Reynolds number is $\sim 10^{12}$ and which will have such structures on scales from 10^{-8}m upward. The interaction between the fastest molecules and these vortical structures is nonlinear, sustaining both against dissipative thermalization. This provides perspective for the second diagram, the European time series of tropospheric ozone mixing ratio from the 1870s to the 1990s (Figure 5.11) at stations well into the free troposphere by virtue of their montane sites, although still near the surface. The increase over the 120 plus years is at least a factor of two (Volz and Kley 1988; Harris et al. 1997) and possibly a factor of five (Marenco et al 1994). There are of course no reliable nineteenth century data on the ozone mixing ratio through the depth of the troposphere. The consequence of this increase would be an increase of intermittency in tropospheric temperature since the nineteenth century, with possible effects on the shapes of the infrared spectral line shapes of water vapour, ozone, and carbon dioxide. Some experiments to examine such line shapes in the presence and absence of photodissociating ozone would make an interesting laboratory study. We note that the tropospheric value of $J[O_3]$ would also be enhanced by the halogen-induced stratospheric ozone loss, which increases J by decreasing the overhead ozone column, see for example Pyle et al. (2005); Gauss et al. (2006). All commonly used thermometers have sufficient inertia that they will take an average of the translational energy of the air molecules and of the occupied rotational energy levels. However, the distribution of these velocities will be different than it was 120 years ago, and in molecular terms we are not dealing with the same conditions now as then. They are different, in different ways, in both the troposphere and in the stratosphere. The molecular state, we have argued, determines the atmosphere's transmissivity to infrared radiation. Harries (1997) showed that small spectroscopic effects, particularly in water vapour, could have very significant effects in the calculation of the radiative balance and hence upon estimates of global warming under greenhouse gas increases. In the next section, we will see that the molecular state can also affect the chemical reactivity.

The retrieval of global fields of molecular species in the atmosphere from spectroscopic instruments mounted on orbiting satellites necessarily involves knowledge of the spectral characteristics of the molecular lines employed. In principle, the effects of overpopulations of fast molecules relative to the thermalized Maxwellian distributions widely assumed in the retrieval algorithms should be detectable. Given the effects of the turbulence on the pressure and temperature fields, one way to proceed would be to take the autocorrelation function of the detected radiance and Fourier transform it to obtain the spectrum. It would be interesting to see if multifractality was present. Even without these turbulent effects, consideration of the formulae for the energy levels in rotating and vibrating molecules together with the complicated overlaps arising from three major and several minor radiatively active molecules suggests that this is a possibility. Statistical simplifications so based might be better than purely random assumptions when trying to economize on the computational cost of line-by-line calculations for the whole atmospheric spectrum.

6.2 Chemical kinetic implications

The rates at which chemical reactions are observed to occur in the atmosphere can be affected by an overpopulation of fast molecules in the speed PDF in two ways. One is on the molecular scale, where there will be more molecules with sufficient translational energy to overcome activation barriers in bimolecular reactions, and fewer with low enough translational energy to participate in atom and free radical recombination reactions. This phenomenon will have to be tackled by laboratory experiment and theoretical chemistry calculations. The second way is on a larger scale, where the vorticity structures associated interactively with the high speed molecules are the mechanism bringing the fluctuations in the reactant concentrations into contact with each other. These will not have the same effect mathematically as the true, random molecular diffusion in three dimensions assumed to underlie the law of mass action in, say, laboratory reactors. This second effect will be evident in the atmosphere, both observationally and during simulation by numerical models.

There was one field mission, by the ER-2 to the Arctic vortex in January–March 2000, where there were some chemical measurements of good enough quality and quantity to sustain an analysis by generalized scale invariance, albeit restricted to $H_1(\mathrm{ClO})$ and $H_1(\mathrm{NO_y})$ by virtue of otherwise essential calibration gaps in the data records (Tuck et al. 2003a). The ozone data traces alone have sufficient continuity, signal-to-noise ratio and time response to sustain such analysis for C_1 and α in addition to H_1.

Molecular velocity is a fundamental quantity in the study of chemical kinetics. We will be very brief here, seeking only to make points relevant to the understanding and modelling of atmospheric chemistry, with particular reference to polar stratospheric ozone loss, where reaction rates and accumulated depletions are large enough for a clear picture to emerge. Modern molecular dynamics of chemical change is a highly developed subject, using a quantum mechanical framework and very refined experimental techniques. An excellent account may be found in Berry, Rice, and Ross (2002c). For a clear, lucidly expressed account of the basic principles, the two earlier books by Hinshelwood (1940, 1951) are recommended, particularly Chapter III in the older book.

The Law of Mass Action says that the rate of a chemical reaction is proportional to the product of the concentrations of the reacting molecules.

The constant of proportionality, the rate coefficient k_i for the elementary reaction between molecules A and B, is measured in a well-stirred reactor in which true diffusion is the only transport process; the reactants have random access to the entire three-dimensional Euclidean space and to each other. The scale of such reaction vessels is typically centimetres to a metre. There is an inherent problem in atmospheric chemistry in using k with either measured or calculated concentrations, arising from the fluctuations in chemical concentrations caused by turbulent wind systems (Tuck 1979; Edouard et al. 1996; Tan et al. 1998; Searle et al. 1998; Tuck et al. 2003a). True diffusion on the scale of a day—ozone loss in the lower stratospheric polar springtime vortices is estimated to be in the range 1–4% day $^{-1}$ (Jones et al. 1989; Rosenlof et al. 1997; Richard et al. 2001)—has been estimated to be effective on scales of tens of centimetres (Austin et al. 1987), but a $10^3 cm^3$ volume will never be undistorted by turbulent vorticity structures for as long as a day, given the evidence for scale invariance and the high Reynolds number in the free atmosphere. Measurements from aircraft (Anderson et al. 1989) or satellites (Waters et al. 1993) are averages over horizontal length scales 4 and 5 orders of magnitude larger than this, respectively. Numerical models have horizontal resolution which usually spans about 2.5 orders of magnitude down from the largest (global) scale. It has been demonstrated that the variability of ozone, wind, and temperature observations from the ER-2 in the lower stratosphere is fractal (Tuck and Hovde 1999; Tuck et al. 1999; Tuck et al. 2002; Tuck et al. 2004) and see also Chapters 4 and 5. We must expect this variability to affect the application of the law of mass action to obtain rates of chemical reaction in the large volumes over which observational techniques average in the atmosphere.

It is apparent from Table 6.1 that ClO and total reactive nitrogen, NO_y, evolved from late January to mid-March; the value of H_1(ClO) decreased from high values of ≈ 0.8 to low values of ≈ 0.33, while $H_1(NO_y)$

Table 6.1 Values of H_z in the Arctic Vortex, January–March 2000; smallest scale 0.029 Hz at 200 ms^{-1}, or 7 km. Bracketed numbers in column 1 refer to the points in Figure 6.2. The ± number is the 95% confidence interval for the associated value of H_z. Time intervals are indicated by flight times in UTC seconds.

Date	Time Interval	H_z(ClO)	$H_z(NO_y[B])$	$H_z(O_3)$	H_z(M)
20000120[1]	37553–47828	0.69 ± 0.13	0.06 ± 0.03	0.34 ± 0.03	0.50±0.05
20000123[2]	31017–38648	0.76 ± 0.16	—	0.30 ± 0.03	0.49 ± 0.05
20000131[3]	38199–43249	0.82 ± 0.11	0.04 ± 0.01	0.24 ± 0.03	0.51 ± 0.05
20000202[4]	35869–53229	0.66 ± 0.15	—	0.36 ± 0.03	0.52 ± 0.07
20000226[5]	30303–43443	0.46 ± 0.05	0.45 ± 0.04	0.32 ± 0.08	0.48 ± 0.07
20000305[6]	35567–39442	0.37 ± 0.20	0.47 ± 0.08	0.34 ± 0.07	0.43 ± 0.04
20000305[7]	52392–57922	0.42 ± 0.17	0.47 ± 0.11	0.33 ± 0.06	0.44 ± 0.07
20000307[8]	28834–43679	0.32 ± 0.08	0.39 ± 0.09	0.37 ± 0.02	0.54 ± 0.07
20000311[9]	46765–52389	0.32 ± 0.07	0.46 ± 0.06	0.39 ± 0.03	0.52 ± 0.08
20000312[10]	37649–48709	0.32 ± 0.07	0.44 ± 0.04	0.36 ± 0.04	0.47 ± 0.06
20000312[11]	51342–58549	0.34 ± 0.03	0.42 ± 0.05	0.34 ± 0.03	0.46 ± 0.06

increased from very low values ≈ 0.05 to values ≈ 0.45. The ozone stayed constant at $H_1(O_3) \approx 0.35$, a value significantly less than that for a passive scalar, $5/9 = 0.56$. A consistent interpretation of these results is that most of the NO_y was in polar stratospheric clouds in late January with concomitant intermittency (spikiness, anticorrelation) arising from high NO_y in the condensed phase interspersed with low concentrations in the gas phase. These clouds rapidly processed the reactive chlorine to produce high, uniform concentrations of ClO, yielding high H_1(ClO). As the PSCs decreased and the sunlight in the vortex increased, the reactive chlorine in the form of Cl and ClO underwent a chemical chain reaction with ozone, resulting in H_1 exponents of the same, non-passive scalar value. Illustrations of sample flight segments of ClO and their scaling are shown in Figures 6.1–6.3; the evolution with time is clear. Figure 6.4 displays the temporal evolution of the scaling exponents of ClO and NO_y, showing their mirror image progression to a common, reactive state. Figure 6.5 shows the steadiness and continuing reaction of ozone with Cl and ClO while the ClO evolves. The multifractal treatment is thus

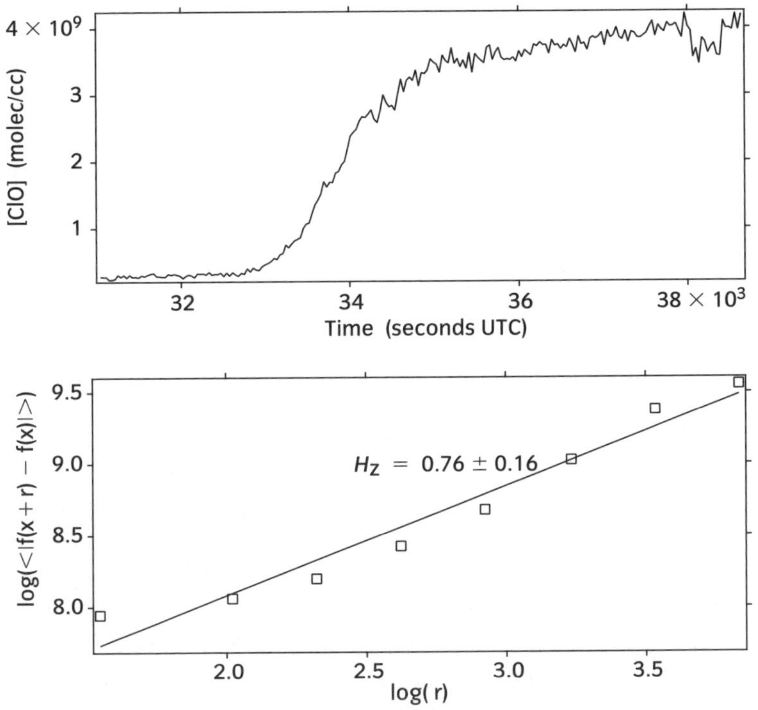

Figure 6.1 The observed ClO trace (upper) and the resulting log-log plot (lower) from which H_z(ClO) was obtained, 20000123. The ER-2 flight took place from Kiruna (68°N, 20°N) during SOLVE, inside the sunlit part of the polar vortex from 33 000 seconds Universal Time, or about 10:20 am local time. The ClO was being produced actively following widespread polar stratospheric clouds; the high value of H_z is a reflection of an effective, recent source.

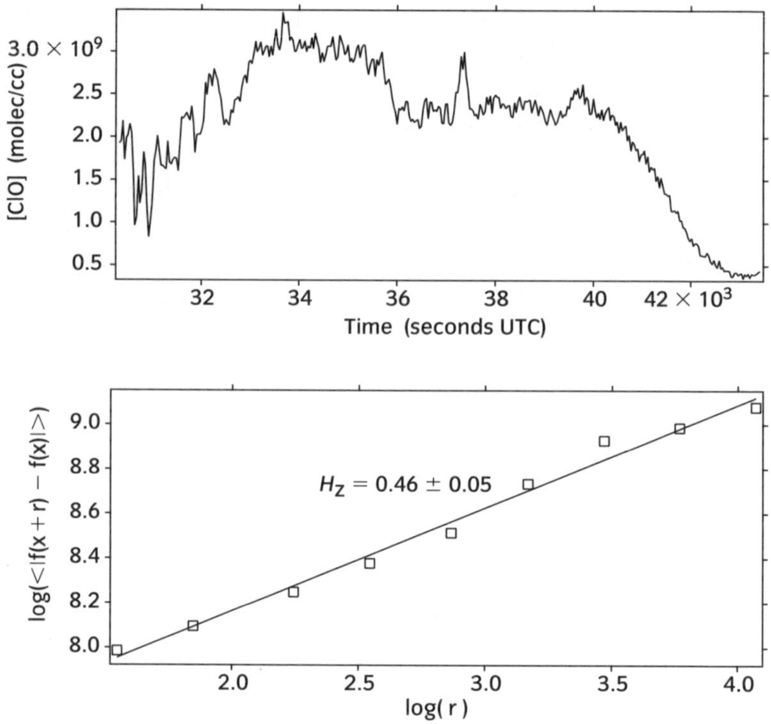

Figure 6.2 As for Figure 6.1, but for the ER-2 flight on 20000226. Note that H_z(ClO) is intermediate between the high value in January and early February (Figure 6.1) and the low value in March (Figure 6.3). The PSC activity was ending by late February and had ceased by mid-March.

consistent, in a numerical model-free way, with the basic physicochemical picture of ozone loss.

A second phenomenon contributing to the lower values of H_1(ClO) in March is that of segregration, where an air parcel with a greater ClO concentration than its neighbours will amplify that difference over time, because of the $[ClO]^2$ squared microscopic dependence and the macroscopic power law dependences >2 (see below). Such segregation would have been enabled in the Arctic vortex in 2000 by the presence of varying amounts of denitrification; reactive chlorine abundance and ozone loss rate were inversely correlated with reactive nitrogen (Gao et al. 2002). This is the reason for the low values of H_1(ClO) in March, which have imposed themselves on $H_1(O_3)$. These ozone exponent values are about the same throughout and lower than either the passive scalar value (0.55˙) or the value of 0.70 seen for the 'end game' in late September over Antarctica, see Figure 4.21, Section 4.3 and Tuck et al., (2002); the relative errors associated with the exponents can be assessed in Figures 4.18–4.21 and 6.2–6.5. $H_1(O_3)$ ~0.35 in late January implies that ozone loss was established before the ER-2 made its first flight in the vortex on 20000120, in agreement with Hoppel et al.

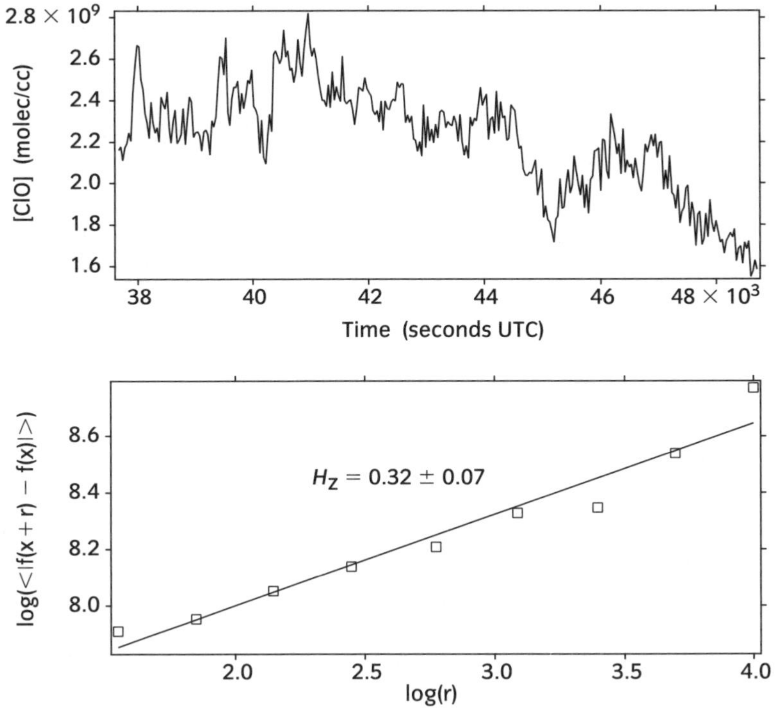

Figure 6.3 As for Figure 6.1, but for the ER-2 flight on 20000312. Note that H_z(ClO) was 0.76 on 20000123, and was 0.32 on 20000312. The value in mid-March was the same as for the ozone and less than the value, 0.56, for the passive scalar nitrous oxide, indicating an active sink for both ClO and for ozone—mutually assured destruction.

(2002). In January, processing by PSCs was ongoing and frequent; provided it was sufficiently efficient, air recently emerged from a PSC should be completely processed (HCl and $ClONO_2$ converted to Cl_2) and a high degree of correlation expected in the reactive chlorine content between neighbouring air parcels. Such an expectation would lead to H_1(ClO) on the high side of the possible zero-to-unity range and would be closer to the condition modelled by Searle et al. (1998). As PSC exposure decreases in frequency from late January to early March, other processes will affect the scaling of ClO. Notably these include turbulent exchange, which in the absence of reaction would make H_1(ClO) tend to 0.55˙. Perfect mixing, by which we mean the attainment of complete anticorrelation among neighbouring intervals on all scales, would make it tend to zero. Such a state is not attainable in a flux-driven, anisotropic gas like the atmosphere. Reaction with ozone, via its chain-carrying partner Cl, will also affect the scaling of ClO. We note that H_1(ClO) in early March is the same as that of ozone, rather than of NO_y(B) which by that time is a chemically inert gas phase tracer.

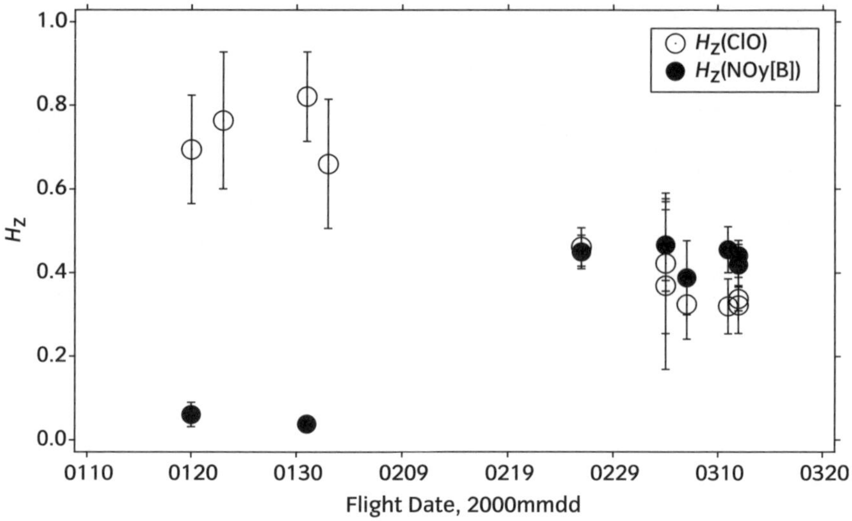

Figure 6.4 H_z(ClO) and H_z(NO_y[B]) versus flight date, for all useable ER-2 flight segments in the Arctic lower stratospheric vortex from Kiruna during SOLVE January to March 2000. The very low values for NO_y reflect the fact that the detected reactive nitrogen is concentrated in single solid PSC particles with very little in the gas phase in flights 1 and 3, whereas on later flights it has values close to but a little less than those for a gas phase tracer, indicating that little is left in the particulate phase. See Figures 6.1.–6.3. for the ClO behaviour.

We therefore suggest that the intimate chemical involvement of ClO and O_3:

$$
\begin{array}{llll}
\mathrm{PSC} \rightarrow \{\mathrm{Cl_2} + h\nu & \rightarrow & \mathrm{Cl} + \mathrm{Cl}\} & \mathrm{R1} \\
2(\mathrm{Cl} + \mathrm{O_3} & \rightarrow & \mathrm{ClO} + \mathrm{O_2}) & \mathrm{R2} \\
\mathrm{ClO} + \mathrm{ClO} + \mathrm{M} & \rightarrow & \mathrm{Cl_2O_2} + \mathrm{M} & \mathrm{R3} \\
\mathrm{Cl_2O_2} + h\nu & \rightarrow & \mathrm{Cl} + \mathrm{ClO_2} & \mathrm{R4} \\
\mathrm{ClO_2} + \mathrm{M} & \rightarrow & \mathrm{Cl} + \mathrm{O_2} + \mathrm{M} & \mathrm{R5} \\
\hline
2\mathrm{O_3} & \rightarrow & 3\mathrm{O_2} &
\end{array}
$$

results in a common scaling exponent, in this case $H_1(\mathrm{ClO}) \approx H_z(\mathrm{O_3}) \approx 0.35$. M is the total concentration of molecules, or the pressure. Note that in the Antarctic inner vortex $H_1(\mathrm{O_3})$ increased from 0.28 to 0.70 during the 8 days after the last PSC exposure, see Figure 4.21 and Section 4.3. It would have been interesting to have seen what would have happened to $H_1(\mathrm{O_3})$ in the Arctic inner vortex in March 2000 in the time for air parcels to complete another PSC-free circuit after the last ER-2 flight on 20000312 (about five days). Recall, however, that the scaling behaviour of ozone shows chemistry and turbulence induced changes in α as well as H_1 (Tuck et al. 2002). There is no contradiction between the low values of $H_1(\mathrm{NO_y[B]})$ and the high values of H_1(ClO) during January shown in Figures 6.1 and 6.4. The NO_y[B] distribution is tending strongly to antipersistence but on the scale of a PSC the exposure of the air to the NO_y particle surfaces and actinic

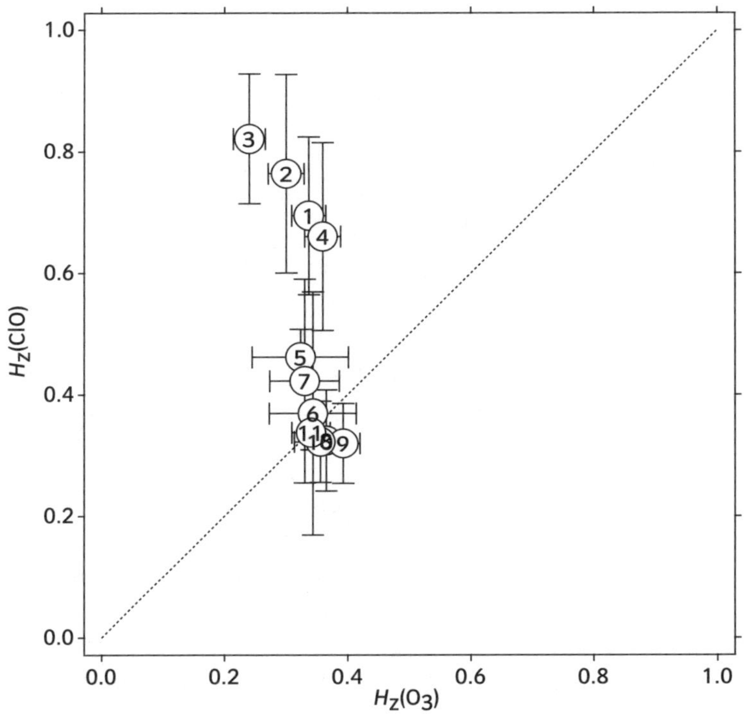

Figure 6.5 H_z(ClO) vs. $H_z(O_3)$, ER-2 data from SOLVE in the Arctic lower stratospheric vortex, January to March 2000. The dotted line is for 1-1 reference only. Note that the scaling of ClO is the same as that of O_3 by early March, and less than the 0.56 expected for an inert scalar. See column 1 of Table 4.3 for numbering of points, and Figures 6.1.–6.3. for examples of the ClO scaling.

insolation results in a ClO distribution tending to persistence. As far as NO_y[B] is concerned, we note that our arguments here mean that the access of HNO_3 vapour to a large, cold nitric acid trihydrate crystal (Fahey et al. 2001; Carslaw et al. 2002) is determined by scale invariant turbulence rather than by molecular diffusion, making it possible to remove the NO_y more quickly from a larger volume of air.

We consider the very simple case in which the rate determining step for ozone loss in the PSC-processed lower stratospheric polar vortex is given by the steady state expression $2k_3[\mathrm{ClO}]^2[\mathrm{M}] = 2J_4[\mathrm{Cl_2O_2}]$, where k_3 is the microscopic rate coefficient in the mechanism of elementary reactions given above, and J_4 is the photodissociation rate of the dimer (Hayman et al. 1986; Molina and Molina 1987; Cox and Hayman 1988; Bloss et al. 2001). We ignore complications arising from bromine chemistry (McElroy et al. 1986; Toohey et al. 1990), the photoisomerization of OClO (Vaida et al. 1989) and other possible reactions of the ClO dimer (Anderson et al. 1989). The objective is to illustrate the principle rather than to perform accurate calculations of ozone loss. The rate expression has [ClO] raised to the power 2 when D, the dimension of the reaction volume, is 3, as is the case in a diffusive regime. Schertzer and Lovejoy (1985) have shown that D for the atmosphere is 23/9, a value of 2.56 which has received recent experimental support from the analysis of ER-2 data (Lovejoy et al. 2001; Tuck et al. 2002, Lovejoy et al. 2004). The value of 23/9 derives from the elliptical nature of atmospheric eddies; the anisotropy arises from the effect of gravity, and is formulated by Schertzer and Lovejoy (1985) as arising from $2 + H_z$, where $H_z = H_h/H_v$ is the ratio of the horizontal and vertical scaling exponents of the wind. The value of $H_h = 1/3$ corresponds to Kolmogorov $k^{-5/3}$ scaling in the horizontal wavenumber spectrum, where k is wavenumber, while $H_v = 3/5$ corresponds to Bolgiano $k^{-11/5}$ scaling in the vertical wavenumber spectrum. As we have seen, measurements along the aircraft trajectory measure H_z because it is not possible to fly it on a truly horizontal flight path to measure H_h, so any observations of H_z (ClO) will refer to chemically induced modifications to the passive scalar value of $0.5\dot{5}$. Note that H_z here is a notation consistent with that of Schertzer and Lovejoy (1985) and Lovejoy et al. (2004). It is called H_1 elsewhere in this book, the first moment evaluation along a flight segment according to Equations (4.8) to (4.10). We are interested in the effects of non-conservative processes, such as chemistry or particle precipitation, on the value of H_1. The value of 5/9 or $0.5\dot{5}$ for passive scalars (tracers) implies an acceleration of the rate of reaction in a space with a lower dimension, a result first noted by Adam and Delbrück (1968) in a molecular biological context; in the limit of denial of a whole dimension to a reacting population it is easily seen that the frequency of encounters among the population increases when it is constrained to a two-dimensional surface rather than to its original three-dimensional volume.

We now investigate whether it is possible to use the ClO scaling behaviour, observed from the ER-2 over approximately 2.5 orders of magnitude in length scale, in the law of mass action equation for reaction $A + B \rightarrow C + D$:

$$-\frac{d[A]}{dt} = -\frac{d[B]}{dt} = \frac{d[C]}{dt} = \frac{d[D]}{dt} = k[A][B] \tag{6.1}$$

or, under the approximations listed above for the reaction mechanism R1-R5,

$$-\frac{d[O_3]}{dt} = 2k_3[ClO]^2[M]. \tag{6.2}$$

The issue is what to use in (Equation 6.2) for the macroscopic case observed by the ER-2 instruments. There are three items, the powers by which [ClO] and [M] are to be raised, and the units and meaning of macroscopic k_{vort} in place of microscopic k_3. To derive these quantities analytically a priori would involve solution of the combined Navier–Stokes equations for atmospheric flow and the differential equations describing the temporal evolution of the chemical reactions R1-R5, from the smallest scales to the largest. The knowledge needed for such an approach is not available. We note that by restricting the problem to two (horizontal) dimensions, and imposing an 'on-or-off', zero or positive condition for the concentrations of ClO and NO_2, Wonhas and Vassilicos (2002) used numerical simulation (contour advection) to deduce a time-dependent fractal dimension for the interface between the ClO-rich air leaving the vortex and the NO_2-rich air into which it was being dispersed, and so calculated the rate of production of $ClONO_2$ in mid latitudes. The essence of our approach is that the observed scaling behaviour of the reactants is neither 2D as it would be under quasi-geostrophic turbulence theory, nor 3D as it would be under isotropic turbulence theory. Instead, it is 23/9 D as argued by Schertzer and Lovejoy (1985), and as observed for the known passive scalar N_2O, in the shape of $H_1(N_2O) \approx 5/9$ (Figure 4.6). It might be possible to arrive at values for k_{vort}, m and n in the rate expression for ozone loss $k_{vort}[ClO]^m[M]^n$ by numerical simulation, but we point out that current state of the art general circulation models cover about 2.5 decades in horizontal scale down from that of a complete great circle (about 44 000 km), whereas about 15 are needed to get down to the scale of vorticity generation by ring currents and the molecular operation of the fluctuation-dissipation theorem. Instead, we proceed by analogy, and show that what determines the rate of reaction in the vortex is the rate of access of [ClO] fluctuations to each other. Compared to a hypothetical volume of air equal to that of the vortex, but in which the fluctuations had access to each other only by a process mimicking true diffusion, the real rate must be accelerated, since the reactants are confined to 23/9D space rather than 3D space. The simplest assumption is that since in 3D space the ozone loss rate is proportional to $[ClO]^2$, then in 23/9D space it will be proportional to [ClO] raised to $2 \times (27/9) \div (23/9)$, or $m = 2.3$ if the chemistry is not fast enough to prevent ClO acting as a passive scalar. This was approximately the case in the flight of 20000226 (Figure 5.11), but not earlier or later. The pressure [M] scales like a tracer, so $H_1(M) = 5/9$ and $n = 1.18$. Recently, Vassilicos (2002) has discussed mixing in turbulent vortices; the difficulty of the 3D problem is acknowledged. Mixing is a word which it is difficult to define precisely in the atmosphere; in my view, it can only be discussed in the light of the fluctuation-dissipation theorem and the unattainability of both smooth, advective flow and of perfect anticorrelation on any scale.

These arguments mean that k_{vort} is an empirical, macroscopic rate coefficient derived from empirical application of the law of mass action to the observed reactant concentrations in the vortex as a function of time, rather in the spirit of the original formulation by Guldberg and Waage in the nineteenth century, before the advent of understanding in terms of atoms, free radicals, and sequences of elementary collisional steps. The power law exponent n for the pressure as measured from the ER-2 is in practice about 1.2, resulting from $H_1(\mathrm{M}) \approx 0.5$ so $n = 3/(2 + 0.5)$. If ClO was a tracer, the value of $H_1(\mathrm{ClO})$ would be $5/9 \approx 0.56$. Its actual average values are 0.73 in late January/ early February and 0.35 in early March. What do they mean? In quantitative terms it means that the rate of ozone loss should be evaluated as $k_{vort}[\mathrm{ClO}]^{2.20}[\mathrm{M}]^{1.2}$ in late January ($2 \times 3/2.73 = 2.20$) and as $k_{vort}[\mathrm{ClO}]^{2.55}[\mathrm{M}]^{1.2}$in early March ($2 \times 3/2.35 = 2.55$). Physically, the ClO is better mixed in March than in late January/early February, and its average scaling exponent at 0.35 is nearer the perfect-mixing limit of 0; the average value of 0.73 in late January/early February is closer to the smooth continuous advection limit of 1. Neither condition, however, corresponds to the completely processed condition modelled by Searle et al. (1998). The intimate and mutually dependent existence of ozone and ClO in the vortex, embodied in the chemical chain reaction mechanism of elementary collisional steps (R1)–(R5), is a kind of *danse macabre*, culminating in the disappearance of both species if the vortex is sufficiently durable, as it is over Antarctica but not the Arctic. The equality of their scaling exponents by early March is a reflection of progression towards this situation. An increase of 10% in the exponent of ClO in Equation 6.2 from 2 to 2.2 can have considerable effects on the computed rate of ozone loss in the inner vortex.

One possible approach to the difficulty of applying the law of mass action in the atmosphere would be to attempt use of the fluctuation-dissipation theorem to arrive at rates of reaction via the fluctuations of chemical number densities over a sufficiently long time (Zwanzig 2001) as discussed in his Chapter 1.4, or by use of a Fokker–Planck equation via the master equation formulation (his Chapter 3.4, pp. 64–66). Such principles have not yet been applied to the very complicated photochemical kinetics determining atmospheric composition, although they have been formulated for the simple recombination of two monomer molecules to dimer; R3 and (Eq. 6.2) suggest that it might work for polar ozone loss. Theoretical arguments have been given, based on using the variance of ClO, that in the event of complete PSC processing the averaging effects in the polar vortex will be minor (Searle et al. 1998). However, the ClO variability was significant on all scales in the case we have considered, particularly after January, signifying variable processing from one air sample to the next. Furthermore, the $1 < \alpha < 2$ condition observed for wind speed, temperature, and ozone would be expected to hold for ClO too, if the data had had sufficient continuity for $\alpha(\mathrm{ClO})$ to be estimable, implying that the variance does not converge.

We note that in the presence of an overpopulated high-speed tail in the molecular PDF, reactions with an energy barrier, that is those accelerated by heating, will be faster in the atmosphere relative to the equivalent container having Maxwellian velocity distributions. On the other hand reactions which are accelerated by cooling, such as free radical recombinations, will be slower. The former category will select from the fast-moving molecules, the latter will select from the slow-moving molecules. It is also true that the rate of ozone photodissociation will be greater in the higher number density regions associated with ring currents and vorticity structures, and slower in their lower number density regions, a reinforcing feedback. Molecular dynamics simulations might cast some light on whether or not such mechanisms actually work in air.

At this juncture we can briefly consider the values of H_1 for the different molecular species which are available. The values can be categorized according to whether they are less than, equal to, or greater than the passive scalar (tracer) value of 5/9, or 0.56. Nitrous oxide, a known tracer in the lower stratosphere, had a mean value of 0.56 during ASHOE/MAESA, and there were some instances where ozone during AASE had this passive scalar value. Total water had it for some flights during AAOE on the DC-8, and for some WB57F flights during WAM and ACCENT. For values less than 0.56, we note occurrences in ClO during middle and late SOLVE, in some ozone during AAOE, AASE, ASHOE/MAESA, and SOLVE, for some total water in AAOE, WAM, and ACCENT, for NO_y in early SOLVE. There are only two instances of values greater than 0.56, ClO in early SOLVE and ozone on the last flight of AAOE, on 19870922. Detailed examination of these results suggests that when there is a sink removing a constituent from the gas phase, the value of H_1 is less than the tracer value, when it is a passive scalar it is equal to it, and when there is a source H_1 is greater than 0.56. The sole high ozone value was accompanied by a high value of $\alpha(O_3)$, the Lévy exponent (Figure 4.21) and a full multifractal, three-exponent theoretical interpretation is not available at present. However, it is possible that under a very low ozone column, continual sunlight and with there having been no exposure to PSCs for several days, the ozone was starting its photochemical recovery, that is, there was a source, an idea that has independent observational support (Russell et al. 1993). The high ClO value early in SOLVE is coincident with a source, from PSC processing. Total water has a sink (low H_1) whenever ice crystal sedimentation under gravity occurs; NO_y was experiencing a gas phase sink via sequestration on to PSC particles, accompanied by some sedimentation, early in SOLVE; and ozone was being lost when it had low values in the polar vortices, as was ClO when its H_1 had declined to values < 0.56 during late SOLVE. It is not obvious why source, tracer, and sink behaviour should be so defined by the H_1 scaling exponent, but it is a result with potential practical utility.

6.3 Cloud physical implications

If there is a non-Maxwellian distribution of molecular speeds in the atmosphere, some of the smallest scale effects will be upon the vapour pressure over both liquid water and ice. The long tail of high-speed molecules will accelerate the access of water vapour molecules from the gas phase to the surfaces of droplets and crystals, compared to the rate that would be attained by simple Einstein–Smoluchowski diffusion. In turn this means that the Clausius–Clapeyron equation will be less precise, and that supersaturation will be harder to define. The question of defining a temperature for the condensed phases, particularly at their surface, becomes more complicated. The long warm tail in the temperature distribution, as seen in observations, will also accelerate the evaporation of small droplets and crystals in cold regions, such as the winter vortex and the tropical tropopause while the long cool tail will decelerate their evaporation in warm regions. We note that laboratory studies of the isotopic composition of water resulting from free molecular evaporation in the absence of condensation show large deviations in the deuterium content from equilibrium values expected from the composition of the bulk liquid, a result demonstrating the importance of a kinetic molecular understanding of cloud microphysical processes in the atmosphere in general (Cappa et al. 2005) and of isotope fractionation in particular (Webster and Heymsfield 2003).

The interaction between the turbulent vorticity structures we have associated with generalized scale invariance and hydrometeors (aqueous particles) might be expected to be largest when both vorticity structures and particles are about the same size. Since ring currents, at least in molecular dynamics simulations, appear at about 10 nm length scale, only the very smallest particles might be unaffected by this mechanism, although they will of course be transported by the turbulent wind field almost as though they were gaseous, because of their low inertia. One might intuitively expect particle–particle interactions among droplets, both of liquid water and chemically more diverse aerosols, and ice crystals to be accelerated by the turbulent vorticity structures so as to be fast compared to a gas in which true diffusion was the sole transport mechanism. Such an effect has been seen in computer calculations of the trajectories of water droplets in turbulent velocity fields, via the so-called 'slingshot effect', where a pair of adjacent vortices concentrates droplets in the regions of convergent airflow. Not only is coalescence accelerated because the collision rate is dependent upon the square of the droplet number density, but long-range correlations occur too (Falkovich et al. 2002; Celani et al. 2005; Falkovich et al. 2006), implying the probable operation of positive feedbacks. To gain a quantitative understanding of such phenomena in the real atmosphere will require considerable advances, both theoretical and experimental.

6.4 Summary

In (H_1, C_1, α) exponent space, atmospheric wind speed and temperature inhabit the ranges $0.4 < H_1 < 0.7, 0.02 < C_1 < 0.10, 1 < \alpha < 2$, associated with asymmetric, long-tailed PDFs. The correlation between ozone photodissociation rate and the intermittency of temperature is significant, and consistent with direct observation of temperature PDFs in the same air on either side of the terminator. While it has been central dogma for decades that radiation, dynamics, and chemistry are coupled, the perspective here is that it happens on the smallest time and space scales because the turbulent vorticity field, radiative transfer, and chemistry arise directly from the dynamical behaviour of molecules in a non-equilibrium environment. Macroscopic formulations of these non-equilibrium, non mean field effects remain to be produced; it will mean representing the direct relations of molecular speed with temperature, vorticity, spectral line shapes, and rates of chemical reaction. It will necessarily be stochastic and scale invariant, requiring major research efforts in a wide range of phenomena. However, if we are correct, a properly based statistical representation of molecular behaviour as it affects turbulence, radiation, and chemistry will be necessary for a solid description of the non-equilibrium energy distributions and transformations in the atmosphere. Once available, it could be used in computer models of the atmosphere, particularly those in which the simulations extend for long enough to be without influence from initial conditions or observations that accurate descriptions of the energy states are necessary; climate modelling is an obvious example.

Reading

Berry, R. S., Rice, S. A., and Ross, J. (2002a), *The Structure of Matter,* 2nd Edition, Oxford University Press, Oxford.

Berry, R. S., Rice, S. A., and Ross, J. (2002b), *Matter in Equilibrium,* 2nd Edition, Oxford University Press, Oxford.

Berry, R. S., Rice, S. A., and Ross, J. (2002c), *Physical and Chemical Kinetics,* 2nd Edition, Oxford University Press, Oxford.

Chapman, S. and Cowling, T. G. (1970), *The Mathematical Theory of Non-Uniform Gases,* 3rd Edition, pp. 93–96, 327 Cambridge University Press, Cambridge.

Chen, T-Q. (2003), *A Non-Equilibrium Statistical Mechanics Without the Assumption of Molecular Chaos*, World Scientific, Singapore.

Finlayson-Pitts, B. J. and Pitts, J. N. Jr. (2000), *Chemistry of the Upper and Lower Atmosphere*, Academic Press, San Diego.
Hinshelwood, C. N. (1951), *The Structure of Physical Chemistry*, Oxford University Press, Oxford.
Landau, L. D. and Lifshitz, E. M. (1980), *Statistical Physics, 3rd Edition, Part 1, Course of Theoretical Physics*, Volume 5, Butterworth-Heinemann, Oxford.
Wayne, R. P. (2000), *Chemistry of Atmospheres, 3rd Edition*, Oxford University Press, Oxford.
Zwanzig, R. (2001), *Nonequilibrium Statistical Mechanics*, Oxford University Press, Oxford.

7 Non-Equilibrium Statistical Mechanics

The Earth's atmosphere is far from equilibrium; it is constantly in motion from the combined effects of gravity and planetary rotation, is constantly absorbing and emitting radiation, and hosts ongoing chemical reactions which are ultimately fuelled by solar photons. It has fluxes of material and energy across its boundaries with the planetary surface, both terrestrial and marine, and also emits a continual outward flux of infrared photons to space. The gaseous atmosphere is manifestly a kinetic system, meaning that its evolution must be described by time dependent differential equations. The equations doing this under the continuum fluid approximation are the Navier–Stokes equations, which are not analytically solvable and which support many non-linear instabilities. We have also seen that the generation of turbulence is a fundamentally difficult yet central feature of air motion, originating on the molecular scale. Non-equilibrium statistical mechanics may offer insight into which steady states a system far from equilibrium as a result of fluxes and anisotropies may migrate, without the need for detailed solution of the explicit path between the states. However, it does not seem possible to demonstrate mathematically that such steady states exist for the atmosphere. A physical view of the planet's past and probable future suggests that the past and future evolution of the sun and its outgoing fluxes of energy may mean that the air-water-earth system may never have been or will ever be in a rigorously defined steady state. Also, to the human population, the detailed, time-dependent evolution is what matters in many respects. Nevertheless, non-equilibrium statistical mechanics is a discipline which should be applicable in principle to yield information about approximate steady states. These steady states may as a practical matter be definable from the observational record, for example the ice ages and the intervening periods evident in the geological record, or between states with two differing global average abundances of a radiatively active gas such as carbon dioxide.

There has been great progress recently in non-equilibrium statistical mechanics, stemming from recent work on the concept of the maximization of entropy production.

7.1 Maximization of entropy production

The maximization of entropy production is a controversial subject. Some recent work has been organized into a set of chapters and collected into a book (Kleidon and Lorenz 2005), which covers a wide range of macroscopic topics, mainly connected with Earth and its subsystems but with some attention paid to larger systems, such as the universe. How useful a global constraint such as maximization of entropy production will be in meteorology is open to investigation; in principle it is applicable to any physical object or set of objects which can be defined as an open system in the thermodynamic sense. The Earth, particularly as regards its surface and fluid envelope, is such a system of course.

Chen (2003) has written a mathematical derivation of turbulent fluid mechanics from the Gibbs distribution governing molecular populations, dispensing with the assumption of molecular chaos. The prose sections of the book say some important things; if the mathematics withstands detailed scrutiny, it will be a major advance. In particular, from what is said about the properties of his 'turbulent Gibbs distribution', it is possible that scale invariance may arise naturally (p. 341), although Chen does not make this claim. To the present author's knowledge, no link has been established yet between the emergence of scale invariance and the production of entropy in an open, flux-driven system like the atmosphere.

Non-equilibrium statistical mechanics, like its equilibrium version, was historically largely the preserve of physical chemists and statistical physicists and so tended to focus on systems of atoms and molecules. In the Nobel Prize-winning work of Onsager and of Prigogine, which was in this tradition, a minimization of entropy production principle was discovered. In contrast to the maximization principle, it is limited to linear single state systems close to equilibrium; these are of no relevance to the Earth's atmosphere.

The general maximization of entropy principle was expounded by Jaynes (1957a, b; 1965; 2003). The fluctuation theorem has been reviewed by Evans and Searles (2002) and the whole approach has been put on a quantitative basis by Dewar (2003, 2005a, b). Critical examinations may be found in Dougherty (1994) and in Balescu (1997).

Here we will pose some questions via a particular example of immediate interest in the context of a molecular view of meteorology, prompted by the discovery of 'ring currents' (vortices) in molecular dynamics simulations of the application of an anisotropy to a population of equilibrated Maxwellian molecules. Consider an air element of such molecules at 200 K temperature, arbitrarily subject on one boundary to a flux of molecules with the same velocity distribution, but with 130 m s^{-1} added to each. We note that such fluid velocities at these temperatures have been observed in the winter sub tropical jet stream and polar night vortex. How will the molecular motion evolve? This can only be 'solved' by a molecular dynamics simulation on a large, fast computer; the current state of the art could handle $\sim 10^{10}$ Maxwellian molecules for $\sim 10^{-8}$ seconds. Despite the smallness of these scales, such an exercise could be extremely informative: would the vortices produced yield a scale invariant structure, and would the scaling exponents be what are observed in jet streams at length scales 8 to 13 orders of magnitude larger? If this were to be the case, the exponents would provide a physically a priori stochastic basis for representing the smaller scales in numerical models of the atmosphere. One can be sure that the meteorological approximation, that the Maxwellian population would continue its true diffusion while being advected at $130\,\mathrm{m\,s^{-1}}$, would not hold. Vorticity structures in the form of 'ring currents' would preclude it. If scaling exponents such as the three in generalized scale invariance were found in such a molecular dynamics calculation, it would be a major advance, particularly if the magnitudes were similar to those observed on scales from 40 to 7×10^6 m in the atmosphere. The exponents would embody a statistical expression of the behaviour of air molecules which could be used in larger scale numerical models of the atmosphere, given the need for the latter to represent the smaller scale processes better than they do currently.

Is there anything that a maximum entropy approach via the fluctuation theorem tells us? The basic physical point is that in a flux (anisotropy) driven situation like our example, the complete domination of either perfectly correlated molecular motion (fluctuation) or completely decorrelated molecular motion (dissipation) would not maximize the entropy production at some steady state between the extreme cases of laminar flow and thermalization (absence of gradients). This is the fundamental root of the competition between organized and random motion, which spreads across many fields and phenomena, commonly expressed in meteorology as 'advection and diffusion'. They are not independent of each other. We note here that so-called Lagrangian air trajectories calculated from gridded meteorological analyses are not really the trajectories of the centre of gravity of a defined mass of air; rather, they are Lagrangian sampling of an Eulerian wind field. When the same algorithm employed for the wind components is used to calculate the mixing ratio of a conserved tracer from the gridded analysis, its mixing ratio evolves in regions of wind shear, for example (Allam and Tuck 1984). The consequences of dissipation cannot be avoided, either in the real atmosphere or in numerical simulations of it, as is immediately apparent from inspection of the wind and shear vectors in Figure 2.3. Co-existent regions of order and dissipation far from equilibrium produce more entropy than either a purely dissipative population, or on the other asymptote, a completely ordered unidirectional flux. In our example, aided by the rapid generation of 'ring currents' in molecular dynamics simulations, the ring currents represent the ordered motion, the rapid equilibration near the most probable velocity represents the dissipation. This is highly reminiscent of our discussion of generalized scale invariance in the atmosphere; there is even a common formalism between thermodynamic inverse temperature $1/k_B T$ and our variable q in Chapter 4, with the partition function being $e^{-K(q)}$ and the free energy being $K(q)/q$ (Chapter 4, Section 3.2, in Schertzer and Lovejoy 1991). It will be interesting to see if a connection can be developed for these generalized scale invariant quantities with the non-extensive entropy developed by Tsallis (2004), which has been linked to scale invariance (Tsallis et al. 2005).

Dewar's formulation says that entropy production, σ_Γ, along a forward path in phase space Γ is governed by the same, Gibbsian, distribution that Jaynes applied to the equilibrium case, namely

$$S_\Gamma = -\sum_\Gamma p_\Gamma \ln p_\Gamma, \tag{7.1}$$

where S_Γ is equal to the logarithm of number of phase space paths Γ with probability p_Γ, where

$$p_\Gamma \propto \exp\left(\frac{\tau\,\sigma_\Gamma}{2k_B}\right) \tag{7.2}$$

Instead of counting microstates as in the equilibrium case, paths are counted in phase space for the non-equilibrium case. From microscopic reversibility the dynamical equations are time-reversible (the dynamical equations are time-symmetric), replacement of Γ by its reversal Γ_R will result in $\sigma_{\Gamma_R} = -\sigma_\Gamma$, so

$$p_{\Gamma_R} = \exp\left(-\frac{\tau\,\sigma_\Gamma}{2k_B}\right) \tag{7.3}$$

from which it follows that the ratio of the probability of the forward path to that of the reversed path is

$$\frac{p_\Gamma}{p_{\Gamma_R}} = \exp\left(\frac{\tau\,\sigma_\Gamma}{k_B}\right) \tag{7.4}$$

This states that the probability of the forward path Γ is exponentially greater than that of the reverse path Γ_R, a proposition that is also truer the longer is the time τ. Violations of the second law of thermodynamics are possible in fluctuations, but not for long or over a big phase space volume, since entropy is extensive.

Figure 7.1 shows a molecular dynamics simulation and the resultant PDF for the strain component of the pressure tensor, with entropy production of both signs (Evans and Searles 2002). The fast molecules in the tail, a minority, produce order while the majority of molecules near the most

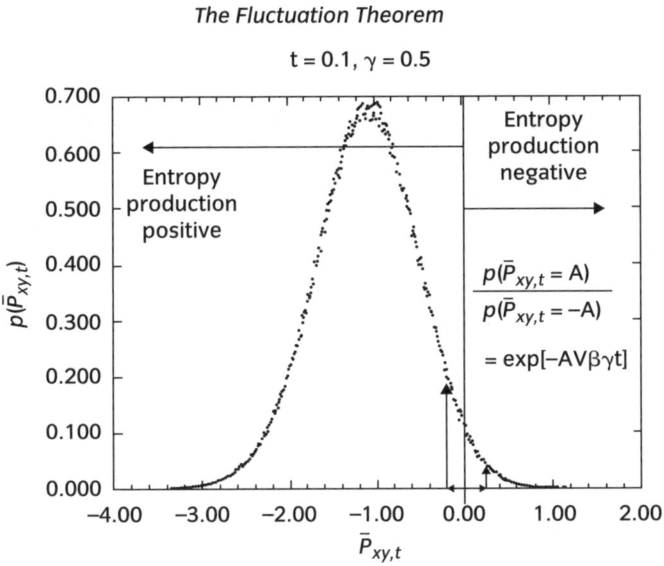

Figure 7.1 How the faster molecules produce 'order' while the slower, more probable ones produce 'disorder.' This is from Evans and Searles (2002) and is the expectation from sheared flow in a molecular gas under a constant strain rate in the x-direction $\gamma = \partial u_x/\partial y$; the system is at constant volume and at constant temperature, T. Temporal averages of the xy-element of the pressure tensor, $\langle P_{xy,t}\rangle$, are proportional to minus the time-average of the entropy production. The dots comprise the histogram of the probabilities for $\langle P_{xy,t}\rangle$; the ratio of the positive to negative probabilities, p, declines exponentially with volume, strain rate and time; negative entropy production is local and short-lived. Thus although the production of 'order' in fluctuations with negative entropy production is small, it is not zero. Note that the emergence of vortices in Figure 8 (Alder and Wainwright 1970) is an example of fluid flow ('ordered') emerging as the result of a mutually sustaining feedback between the faster molecules and the 'ring currents'. The real atmosphere is not under a constant strain rate, is not at constant temperature, and experiences anisotropies from gravity, planetary rotation, the solar beam, and the planetary surface. Fluid flow in the atmosphere emerges via the 'ring current' mechanism as translationally hot ozone photofragments recoil not into a thermalized bath but into a pre-existing set of scale invariant vorticity structures from 10^{-8} to 10^{7} metres. In the real atmosphere the histogram would therefore be expected to have a non-Gaussian shape, with a power law tail rather than exponential one at negative entropy productions. The peak values still correspond to dissipation (positive entropy production) and permit operational definition of atmospheric temperature.

probable velocity produce disorder. A similar figure could be drawn for the fat-tailed PDF of molecular speeds in a molecular dynamics simulation with anisotropy producing 'ring currents', vortices. The tail toward zero and positive values on the abscissa would be power law rather than exponential, because of the 'ring current' mechanism. Recall the remarks in Chapter 3, concerning the contrast between the molecular dynamics-non equilibrium statistical mechanics view of the origins of order and dissipation with the classical meteorological, Langevin-based resolution of a fluid variable into an 'organized' mean and the 'disorganized' departures, eddies, from it.

The stable, random non-Gaussian processes considered by Lévy produce probability density functions characterized by $S_\alpha(\sigma, \beta, \mu)$ where α is the stability index we have used in our formulation of generalized scale invariance, σ is scale factor (standard deviation for a Gaussian), β is skewness and μ is the mean (Samorodinsky and Taqqu 1994). There are only three known cases that can be stated in closed form:

Gaussian: $S_2(\sigma, \beta, \mu)$ with probability density $\frac{1}{2\sigma\sqrt{\pi}} \exp\left(-\frac{(x-\mu)^2}{4\sigma^2}\right)$,

Cauchy: $S_1(\sigma, \beta, \mu)$ with probability density $\frac{\sigma}{\pi\left((x-\mu)^2+\sigma^2\right)}$,

Lévy: $S_{1/2}(\sigma, \beta, \mu)$ with probability density

$$\sqrt{\frac{\sigma}{2\pi}} \frac{1}{(x-\mu)^{3/2}} \exp\left(-\frac{\sigma}{2(x-\mu)}\right).$$

For $S_\alpha(\sigma, \beta, \mu)$ the upper and lower tails of the PDFs decrease like a power function. The rate of fall-off depends on α; the smaller is α the slower is the decay and the fatter is the tail. When $\alpha < 2$ the distributions have infinite variance and when $\alpha \leq 1$ the mean is infinite too. Fortunately we know empirically that for atmospheric variables $1 < \alpha < 2$. The full set of conditions governing $S_\alpha(\sigma, \beta, \mu)$ is $\alpha \in [0, 2]$, $\sigma \geq 0$, $\beta \in [-1, 1]$ and $\mu \in \mathrm{R}^1$. β is definable in terms of statistical moments as $\left|\Delta x^3\right| / \left|\langle \Delta x \rangle^2\right|^{3/2}$, the skewness.

It is not immediately clear how to relate H_1 and C_1 from generalized scale invariance to σ, β, and μ. But we know $\alpha \approx 1.6$ for our atmospheric data, and we can calculate σ, β, and μ. The atmosphere qualifies as a non-Gaussian, Lévy stable random process. In turn this has implications for prediction and numerical modelling; Gaussians may not be useful, while intermittency and multifractality have to be recognized.

Finally, we note for completeness that very recently Tsallis et al. (2005) have examined the relationship between the scale invariant occupancy of phase space and extensiveness (i.e. addivity) of entropy. Any connection between entropy production and scale invariance would be of great interest, but is as yet apparently far from being applicable to a real system like the atmosphere, with its large variety of different subsystems and gigantic number of degrees of freedom, arising from $\sim 8 \times 10^{43}$ molecules. An introduction to this topic may be found in Tsallis (2004).

7.2 Summary

Non-equilibrium statistical mechanics shows that the continual interplay between fluctuation and dissipation has molecular roots and that it can be formulated for any macroscopic system. A complete description of such a system must involve quantum mechanical formulations of the Liouville equation, with intermolecular potentials that have both attractive and repulsive components. One could hope that molecular dynamical simulations of air, using Maxwellian molecules, on the largest and fastest computers would yield scale invariance with similar exponents to those observed in the atmosphere on scales 8 to 13 orders of magnitude larger. If this should happen, a way forward to macroscopic simulation might be possible. It is however by no means certain that the macroscopic Earth system can be formulated in a non-equilibrium statistical framework that will have the necessary steady states, or that the basic dynamics can be expressed in a computationally affordable manner. The technique has developed sufficiently fast recently however, and has shown enough promise, that the effort should be well worth pursuing.

In an atmospheric context, the inherently statistical basis implies that analysis in terms of advection only is limited, as can be seen directly from the observed winds and shear vectors in Figure 2.3, and as is further implied by the molecular scale 'ring current' generation of vorticity. Generalized scale invariance applied to atmospheric observations locates them at $H_1 = 0.56$ in an exponent space $H_1 \in [0, 1]$ with zero being complete decorrelation and unity being complete correlation. $H_1 = 0.50$ is 'randomness', that is no correlation or decorrelation; the atmosphere is thus on the organized side of randomness.

One consequence of these results is that there is no reversibility in atmospheric motion, a result consistent with the maximum entropy production view from non-equilibrium statistical mechanics and with observations (Shapiro 1980; Tuck et al. 2004).

The increment of 0.06 in the canonical datum might not seem like much, but individual realizations in strong jet streams have $H_1(s)$ approaching 0.70 in the horizontal and show significant correlation with the magnitude of the horizontal shear (Figure 4.9). In the vertical in such jet streams, $H_1(s)$ approaches 0.90 and shows significant correlation with the vertical shear of the horizontal wind (Figure 4.10). In the vertical, the effect of gravity on temperature through the hydrostatic relation is dominant, although the horizontal wind and the humidity profiles are much less coherent than that of temperature. These correlations are the signature of the emergence of order, in the shape of a recognizable flow in jet streams, much as 'ring currents' emerge from a microsopic anisotropy in the form of a flux. The larger scales respond to planetary rotation, the solar flux and the surface topography. These phenomena are probably why weather forecasting works, but the fact that the H_1 exponent is less than unity limits the success in space and time.

8 Summary, *Quo Vadimus?* and Quotations

In this chapter, we offer a summary of the book's results and conclusions, ask what future developments might be contemplated, both theoretical and experimental, and provide some scientific quotations which seemed relevant. The quotations are collected here rather than dispersed through the text, because some of them apply at several junctures and one or two apply to the whole book. It is hoped that they will underline some important points in a memorable and even entertaining manner.

8.1 Summary

Application of generalized scale invariance to large amounts of research quality in situ airborne observations of the free troposphere and lower stratosphere has shown that the atmosphere behaves as a random, non-Gaussian, Lévy stable process. The scaling exponents describing the resultant statistical multifractality are the conservation H_1, the intermittency C_1 and the departure from monofractality α, the Lévy exponent. They had average values of 0.55, 0.05, and 1.6 respectively as deduced from airborne time series of wind speed and temperature. Certain regimes, such as jet streams, however showed correlation within the mean; the value of $H_1(s)$ for horizontal wind speed s was positively correlated with the magnitude of the horizontal speed shear and the value of $H_1(T)$ for temperature was positively correlated with the meridional (equator-to-pole) temperature gradient. The value of $H_1(s)$ in the vertical showed clear correlation with vertical measures of jet stream strength, such as depth and maximum speed. The vertical scaling of temperature showed the paramount influence of gravity, having H_1 close to unity, while horizontal wind speed and relative humidity were about 0.75. These results show that large scale ordered flow can be interpreted as emerging from less ordered smaller scale motions. At the same time, the smaller scale motions are never truly random in the atmosphere and the larger scale motions are never perfectly correlated, smooth flow.

Ozone and water, while occasionally behaving as passive scalars, that is to say as tracers, more often showed the presence of sources and sinks: a numerical model-independent demonstration of the operation of photochemistry and precipitation respectively. In cold regions, such as the lower stratospheric Arctic winter vortex and the tropical tropopause, the temperature PDF had a cold most probable value and a long warm tail, see Figure 4.1. In the warm summertime anticyclone in the Arctic lower stratosphere, however, the reverse behaviour was observed: a warm most probable value with a long cool tail. Turbulent heat flux is central. A surprising result was that the intermittency of temperature in the Arctic lower stratosphere over winter and summer showed positive correlation with the observed ozone photodissociation rate. Independent flights either side of the Arctic terminator in the same air mass confirmed that the direct solar heating was large enough to support such an effect.

We argue that in order to explain the dependence of temperature intermittency on ozone photodissociation rate, it is necessary to appeal to a mechanism whereby the translationally hot atomic and molecular oxygen photofragments are not thermalized in a few collisions to produce local thermal equilibrium. Such a mechanism can be found in molecular dynamics simulations of equilibrated Maxwellian molecules subjected to a dis-equilibrating anisotropy, such as a flux. The mutually sustaining interaction between the resultant ‘ring currents’ (vortices) and the overpopulated tail of high-speed molecules is the key to the molecular generation of vorticity and turbulence. Thus the intermittent vorticity structures represent the emergence of ordered flow from random molecular motion and contain the lower-entropy, higher-speed molecules. The dissipation occurs among the numerous, lower-speed, higher-entropy molecules in the most probable parts of the speed PDF, where entropy is produced via the rapid thermalization of speed in a few collisions and a measurable temperature is effectively maintained. Such a picture is the reverse of vorticity structures representing dissipation of a more ordered, larger scale advective flow. It is also contrary to the classical approach of considering the mean to represent organization in a flow, and the eddy departures from it to represent disorder. We note that the atmosphere always operates in the presence of anisotropies: the solar flux, gravity, planetary rotation, and the surface topography. Additionally, any atmospheric volume, of any scale, will experience fluxes of air molecules in a manner consistent with the observed scale invariance. Accordingly, thermalization is never complete on any scale and local thermodynamic equilibrium does not obtain.

The energy is input to the atmosphere by molecules absorbing photons; the energy must propagate upscale from this smallest of scales. The propagation to larger scales must involve molecules moving, and will involve the overpopulation of faster molecules in larger and larger vorticity structures, up to and including jet streams with core wind speeds that can be a significant fraction of the molecular speed, which will have been shaped and influenced by the effects of the boundaries at the sea surface, the ice, the land surface and its vegetative cover. All these surfaces exhibit fractality and scale invariance. The turbulent, scale-invariant vorticity structures will impose themselves on the number densities of the radiatively active gases and particles, via the wind field, and thence to the radiation field. Energy is thus continually input on all scales and consequently the notion of a conservative energy cascade from the largest scale to the dissipative scale is of limited utility. Through the molecular origins of vorticity and turbulence, the chemical and radiative properties of the atmosphere will also be influenced, via the effects of molecular speed in reactive collisions and via the effects of translational speed in the collisional determination of spectroscopic line shapes in the rotational and vibrational fine structure of water vapour, carbon dioxide, ozone, and methane. It is known from laboratory experiment for example that the temperature dependence of the broadening coefficient of studied water vapour rotational lines depends upon the rotational quantum number in a way indicating that the line shape, and hence the infrared absorption and emission, is dependent upon the relative velocity of the colliding molecules. In turn, a long-tailed molecular speed distribution will increase the effect, and do so in the line wings, where there is less self-absorption and hence more leverage via radiative response to greenhouse gas increases. An accurate description of the energy distribution in the atmosphere and hence of air temperature must encompass an adequate statistical physics at the molecular level, including the generation of vorticity via 'ring currents' and the connection to the scale invariance of wind, temperature, and composition. This is certainly not the case in current atmospheric models.

8.2 *Quo vadimus?*

Some experimental tests of what we have proposed as the mechanism for the molecular generation of vorticity, its upscale propagation, and its effects on the atmospheric state can be envisaged. Similarly, theoretical molecular dynamics simulations of 'air' on large, fast computers could prove to be important in the emergence of turbulent fluid mechanics from flux-driven molecular populations.

First and foremost, high resolution, high quality in situ measurements are required throughout the global atmosphere, to ascertain observationally whether the scaling laws seen from manned aircraft primarily investigating stratospheric composition actually obtain for the meteorological variables and the radiatively active gases, particles, and clouds which determine climate. A plan and the means to do this, vertically and horizontally, actually exist and are achievable with current technology (MacDonald 2005).

A better characterization of the actual motion of the autonomous aircraft and dropsondes through the air would need to be built in from first principles, so that it would be possible to calculate the motion of the platform as a problem in Newtonian physics. The data must be good enough to support the determination of all three scaling exponents for as many measured variables as possible, but minimally including winds, temperature, pressure, water, a passive scalar (tracer), and any chemically active species of interest. In practice, this means combining high frequency and no data drop-outs with low random error over long sampling paths. Observations of $J[O_3]$ would add particular insight into the intermittency of temperature.

Laboratory experiments to investigate the high-resolution spectroscopy of the rovibrational lines of water vapour, carbon dioxide, methane, and ozone itself as a function of temperature and pressure in the presence and absence of ozone photodissociation appear to be eminently feasible. An equivalent experiment could be done in the atmosphere by using an open cell laser instrument, in the free air away from the turbulence caused by the presence of the platform, to study the simultaneous behaviour of well-characterized individual water vapour, carbon dioxide, and methane lines as a function of pressure and temperature. Simultaneous measurement of water vapour, temperature, and pressure might be possible, for comparison with other, independent techniques such as frost point instruments, platinum resistance thermometers, and absolute pressure sensors. Agreement between these spectroscopic and aeromechanical 'thermometers' would be an important constraint. Should there be observable effects caused by translationally hot photofragments of atomic and molecular oxygen, it will be necessary to investigate how the observed increase in free tropospheric ozone by a factor of two to five during the twentieth century has altered the transmission of infrared radiation, and indeed of what we mean by atmospheric temperature. We can measure it with calibrated thermometers of course: can we interpret it correctly?

A much more difficult experiment would be to measure the velocity distribution of air molecules directly, during day and night and in varying conditions of temperature, pressure, humidity, and chemical composition. Even with modern high vacuum techniques, molecular beam velocity selectors and detectors, it would be a very difficult experiment—but also a very informative one.

It would be worth seeing if the rates of chemical reaction in the atmosphere could be obtained by direct observations designed to test the fluctuation-dissipation theorem. For example, it can be formulated for a simple reaction such as the recombination of two monomers to their dimer; such a reaction involving ClO plays a central role in the ozone hole, and the polar vortex is an excellent prototypical system offering the rare luxury of large signals above background variation. The evolution of the lower Antarctic stratospheric ozone loss in late September appears from the ozone scaling exponents to develop to a stage not reached in the Arctic, see Figure 4.21. The Antarctic would therefore be the preferred location for such an experiment.

Current computer capacity has been demonstrated to be capable of molecular dynamics simulation of atom populations as large as 10^{10}, affording scope, for example, to examine vorticity generation by photodissociating ozone molecules in 'air' and how the fluid mechanical behaviour evolves from equilibrium. Any generation of scale invariance would be of great importance, both qualitatively and quantitatively. If the generalized scale invariance exponents existed in these simulations, and better, if they had magnitudes similar to those observed at macroscopic scales in the atmosphere, it would represent fundamental progress in establishing the link between molecular scale 'ring currents' and turbulent atmospheric vorticity.

The shortcomings in model representations of the stratosphere (IPCC, 2001) include a global mean cold bias, incorrectly orientated polar night jet streams—they slope poleward with increasing height instead of equatorward—and very scattered degrees of latitudinal separation between the subtropical and polar night jet streams. It is noticeable that the cold bias could be caused by the effects that have been described in Chapters 5.2 and 6.1. The overpopulation of high-speed molecules, acting through the Alder and Wainwright (1970) 'ring current' mechanism, could greatly affect jet streams, as has been argued, in view of the fact that core jet stream wind speeds can reach $130\,\mathrm{m\,s^{-1}}$ in the upper troposphere and lower stratosphere, a significant fraction of the most probable molecular speed. In the upper stratosphere, $200\,\mathrm{m\,s^{-1}}$ has been analyzed in the austral winter, over half the speed of sound. Could a molecular-based modification of the Navier-Stokes equation be necessary to tackle these shortcomings?

Finally, it would be worth experimenting with a 'bottom-up' approach to sub-grid scale parametrization in numerical models of the atmosphere. If the view is correct that atmospheric vorticity and turbulence emerge from the basic collisional dynamics of the fast molecules in the overpopulated high-speed tails of the PDF, the fundamental physics of the energy flow and distribution must incorporate the smallest scales correctly. The microscopic and macroscopic temperatures must be consistent. Scale invariance is an important constraint. The observed scaling exponents embody a statistical expression of how the atmosphere sustains its energy fluxes. The task would of course be more likely to succeed if the experimental, observational, and molecular dynamical underpinnings outlined above were to be available.

8.3 Some relevant quotations

The following, mostly scientific, quotations have been placed here to enable the reader to associate them with the text at appropriate junctures. Some commentary about location and import is given where appropriate. Several are of wide applicability and are better positioned here than repeated in the text. There are no claims to perfect accuracy. In no particular order:

- '*Does the wind possess a velocity? This question, at first sight foolish, improves on acquaintance.*' [L. F. Richardson].

It is necessary to specify the averaging domain in both time and space to give a meaningful wind, see Figure 2.3. Even so, it should be apparent that this average wind cannot be expected to transport atmospheric species realistically.

- '*Fourier analysis is a clever French trick to impose a symmetry on nature that it doesn't really have.*' [Robert Murgatroyd, Joseph Farman].

The fluctuation-dissipation theorem, long-tailed PDFs, scale invariant turbulence, and the molecular nature of air ultimately render sine and cosine waves inapplicable. In the opinion of the author it is also true that the atmosphere is easier to understand in physical space, even if it is sometimes easier to manipulate mathematical surrogates in Fourier space. In the atmosphere, meteorological 'waves' are limited mainly to a small number of repeat cycles of variable frequency and amplitude with obvious signs of roughness and dissipation.

- '*I am an old man now, and when I die and go to heaven there are two matters on which I hope for enlightenment. One is quantum electrodynamics and the other is the turbulent motion of fluids. About the former I am rather optimistic.*' [Sir Horace Lamb, 1932].

(Later variants have been attributed to von Kármán in Caltech classes and to Heisenberg on his deathbed.)
One can safely say that computer simulation is going to be the way forward, whether working on microscopic or macroscopic scales, given the intractability of the Boltzmann H-equation and the Navier–Stokes equation to analytical solution. Whether one is dealing with air, water or biological fluids Lamb's position still seems justified, particularly in view of the postwar successes with QED.

- '*That's not noise, that's music.*' [Richard Feynman]

A great deal of what is often dismissed or treated as noise on small scales in the atmosphere is in fact not random and is of fundamental importance to transport. The atmosphere is much less amenable to averaging than a typical laboratory experiment, and its internal frequencies are less pleasant to listen to than, say, its transmission of the sound waves from a Stradivarius.

- '*Shut up and calculate.*' [Feynman again, answering student questions about the meaning of quantum mechanics.]

One wonders if climate modellers ever find themselves repeating Feynman's words.

- '*One cannot bury Nernst too often.*' [Max Planck, at the *n*th interment of the discoverer of the 3rd law of thermodynamics, necessitated by shifting political boundaries in central Europe in the first half of the twentieth century.]

Meteorology buries the 3rd law by normalizing to surface pressure. Ultimately, however, an absolute temperature scale with a zero depends upon the existence of countable, discrete quantum mechanical states.

- '*Science proceeds one funeral at a time.*' [Planck again, at Mach's funeral.] (The weak version has 'retirement' instead of 'funeral'.)

Comment seems superfluous; famous examples of being wrong on really big questions include Kelvin on the age of the Sun and Earth, Mach on the reality of atoms and molecules, Rutherford on the utility of nuclear energy, Jeffreys on continental drift, and Einstein on the probabilistic interpretation of quantum mechanics. The notion of a scale gap in atmospheric motion between the synoptic and mesoscales has endured beyond reason.

- '*With thermodynamics, one can calculate almost everything crudely; with kinetic theory, one can calculate fewer things, but more accurately; and with statistical mechanics one can calculate almost nothing exactly.*' [Eugene Wigner]

One might wish to insert 'non-equilibrium' before 'statistical mechanics', but half a century on there is still much truth in these remarks.

- '*Your Majesty, during the course of a long academic career, I have observed that I can become inebriated by imbibing scotch and soda, brandy and soda, rum and soda, and gin and soda, but I have not concluded that soda is intoxicating.*' [Sir Edward Appleton, explaining to the King of Norway his Nobel Address remark that correlation did not prove cause and effect.]

Figures 4.9, 4.10, 4.11, 5.2 and 5.3 contain examples of rather unexpected correlations. The provision of explanations should be done against the background of this quotation. Some philosophers of science might say that correlation is all we have, but the author prefers the notion that causality in the form of molecular collisions producing ring currents is in fact a scientifically useful example of causality, one potentially with predictive power moreover.

- '*More is different.*' [Philip Anderson]

The more volume there is in which matter is contained, or the more matter there is under consideration, the harder it is to ensure uniformity, or alternatively put, the harder it is to prevent anisotropy and emergence of larger scale order and complexity. Perfectly randomized, uniform samples of gas exist only in statistical mechanics textbooks, or are approximated in small volumes in competent laboratories.

- '*To be uncertain is to be uncomfortable, but to be certain is to be ridiculous.*' [Arthur Holmes, quoting Goethe in translation, on receiving the Wollaston Medal of the Geological Society for dating the origin of the Earth via isotope ratios].

Poets have messages for scientists—and so do comedians:

- '*Time flies like an arrow, but fruit flies like a banana.*' [Marx: Groucho, not Karl].

This may serve as a warning of how difficult it is to explain some scientific principles to non-scientists, let alone to get those with ready access to cameras and microphones to take them seriously.

8.4 The arrow of time

It seems to be de rigueur to say something about the fact that time appears to have a direction, in spite of the time symmetry of the dynamical equations. We will be brief. Newton's view that 'time flows equably in and of itself' has been challenged since Loschmidt and then Zermelo raised objections to Boltzmann's use of the H-equation to provide kinetic molecular justification for the second law of thermodynamics. There are instructive arguments in Zwanzig (2001) and in van Kampen (2002) on the level of the behaviour of molecular populations. We note that Poincaré showed that the many-body problem could not be solved, where 'many' means more than two. Time's arrow may be a result of this; we suggest that like politics it is an emergent property of populations numbering three or more. Ultimately of course time and gravity are unified by general relativity. Anisotropy and the existence of large particle populations are ultimately questions for theoretical physics covering a staggering range of scales and phenomena.

References

Aberson, S. D. and J. L. Franklin (1999), Impact on hurricane track and intensity forecasts of GPS dropwindsonde observations from the first season flights of the NOAA Gulfstream-IV jet aircraft, *Bull. Amer. Meteorol. Soc.,* **80**, 421–427.

Adam, G. and Delbrück, M. (1968), Reduction of dimensionality in biological diffusion processes, in *Structural Chemistry and Molecular Biology,* A. Rich and N. Davidson, eds., pp. 198–215, W. H. Freeman, San Francisco.

Alder, B. J. and Wainwright, T. E. (1970), Decay of the velocity autocorrelation function, *Phys. Rev. A*, **1**, 18–21.

Alder, B. J. (2002), Slow dynamics by molecular dynamics, *Physica A*, **315**, 1–4.

Allam, R. J. and Tuck, A. F. (1984), Transport of water vapour in a stratosphere-troposphere general circulation model. II. Trajectories, *Q. J. R. Meteorol. Soc.,* **110**, 357–392.

Anderson, J. G., Brune, W. H., Lloyd, S. A., Toohey, D.W., Sander, S. P., Starr, W. L., Loewenstein, M., and Podolske, J. R. (1989), Kinetics of O_3 destruction by ClO and BrO within the Antarctic vortex: analysis based on in situ ER-2 data, *J. Geophys. Res.*, **94**, 11,480–11,520.

Anderson, J. G. and O. B. Toon, (1993) Airborne Arctic Stratospheric Expedition II: an overview, *Geophys. Res. Lett.,* **20**, 2499–2502.

Austin, J., Pallister, R. C., Pyle, J. A., Tuck, A. F., and Zavody, A. M. (1987), Photochemical model comparisons with LIMS observations in a stratospheric trajectory coordinate system, *Q. J. R. Meteorol. Soc.,* **113**, 361–392.

Baldwin, M. P., Rhines, P. B., Huang, H.-P. and McIntyre, M. E. (2007), Atmospheres: the jet stream conundrum, *Science,* **315**, 467–468.

Balescu, R. (1997), *Statistical Dynamics: Matter out of Equilibrium*, Chapter 16.2D, Imperial College Press, London.

Baloïtcha, E. and Balint-Kurti, G. G. (2005), The theory of the photodissociation of ozone in the Hartley continuum: effect of vibrational excitation and $O(^1D)$ atom velocity distribution, *Phys. Chem. Chem. Phys.,* **7**, 3829–3833.

Bannon, J. K., Frith, R., and Shellard, H. C. (1952), Humidity of the upper troposphere and lower stratosphere over Southern England, Meteorological Office: *Geophysical Memoirs, XI*, No. 88, 36 pp., HMSO, London.

Barenblatt, G. I. (1996), *Scaling, Self-Similarity, and Intermediate Asymptotics*, p. 253, Cambridge University Press, Cambridge.

Batchelor, G. K. (1951), The application of the similarity theory of turbulence to atmospheric diffusion, *Q. J. R. Meteorol. Soc.,* **77**, 315–317.

Batchelor, G. K. and Townsend, A. A. (1949), The nature of turbulent motion at large wavenumbers, *Proc. R. Soc. Lond.* A **199**, 238–255.

Bell, J. B. and Marcus, D. L. (1992), Vorticity intensification and transition to turbulence in the three-dimensional Euler equations, *Commun. Math. Phys.,* **147**, 371–394.

Beltrami, E. (1871), Sui principi fondamentali della idrodinamica, *Mem. Acc. Sci. Bologna,* **1**, 431–476.

Bengtsson, L. (1999), From short-range barotropic modelling to extended-range global weather prediction: a 40-year perspective, *Tellus,* **51A-B**, 13–32.

Berry, R. S., Rice, S. A., and Ross, J. (2002a), *The Structure of Matter,* 2nd Edition, Oxford University Press.

Berry, R. S., Rice, S. A., and Ross, J. (2002b), *Matter in Equilibrium,* 2nd Edition, Oxford University Press.

Berry, R. S., Rice, S. A., and Ross, J. (2002c), *Physical and Chemical Kinetics,* 2nd Edition, Oxford University Press.

Bethan, S., Vaughan, G., and Reid, S. J. (1996), A comparison of ozone and thermal tropopause heights and the impact of tropopause definition on quantifying the ozone content of the troposphere, *Q. J. R. Meteorol. Soc.,* **122**, 929–944.

Bloss, W. J., Nickolaisen, S. L., Salawitch, R. J., Friedl, R. R., and Sander, S. P. (2001), Kinetics of the ClO self-reaction and 210 nm absorption cross section of the ClO dimer, *J. Phys. Chem. A,* **105**, 11,226–11,239.

Bolgiano, R. (1959), Turbulent spectra in a stably stratified atmosphere, *J. Geophys. Res.,* **64**, 2226–2229.

Breene, R. G. (1981), *Theories of Spectral Line Shapes*, John Wiley, New York.

Brewer, A. W. (1944), Work carried out in the Fortress aircraft allotted to H.A.F. for meteorological duties, *MRP* **169**, Meteorological Research Committee, Air Ministry.

Brewer, A. W., Cwilong, B. M., and Dobson, G. M. B. (1948), Measurement of absolute humidity in very dry air, *Proc. Phys. Soc.,* **60**, 52–70.

Brewer, A. W. (1949), Evidence for a world circulation provided by measurements of the helium and water vapour distribution in the stratosphere, *Q. J. R. Meteorol. Soc.,* **75**, 351–363.

Briggs, J. and Roach, W. T. (1963), Aircraft observations near jet streams, *Q. J. R. Meteorol. Soc.,* **89**, 225–247.

Brune, W. H., D. W. Toohey, J. G. Anderson, W. L. Starr, J. F. Vedder, and E. F. Danielsen (1988), In situ northern mid-latitude observations of ClO, O_3 and BrO in the wintertime lower stratosphere, *Science,* **242**, 558–562.

Cappa, C. D., Drisdell, W. S., Smith, J. D., Saykally, R. J., and Cohen, R. C. (2005), Isotope fractionation of water during evaporation without condensation, *J. Phys. Chem. B,* **109**, 24, 391–24,400.

Carslaw, K. S., Kettleborough, J. A., Northway, M. J., Davies, S., Gao, R.-S., Fahey, D. W., Baumgardner, D., Chipperfield, M. P., and Kleinböhl, A. (2002), A vortex-scale simulation of the growth and sedimentation of large nitric acid hydrate particles, *J. Geophys. Res.* **107(D20)**, Art. No. 8300, doi: 10.1029/2001JD000467.

Celani, A., Falkovich, G., Mazzino, A., and Seminara, A. (2005), Droplet condensation in turbulent flows, *Europhys. Lett.,* **70,** 775–781.

Chapman, S. (1916), On the law of distribution of velocities, and on the theory of viscosity and thermal conduction, in a non-uniform simple monatomic gas, *Phil. Trans. Roy. Soc. A,* **216,** 279–348.

Chapman, S. and Cowling, T. G. (1970), *The Mathematical Theory of Non-Uniform Gases*, 3rd Ed., pp. 93–96, 327, Cambridge University Press, Cambridge.

Chen, T. Q. (2003), *A Non-equilibrium Statistical Mechanics Without the Assumption of Molecular Chaos,* World Scientific, Singapore.

Chigirinskaya, Y., Schertzer, D., Lovejoy, S., Lazarev, A., and Ordanovich, A. (1994), Unified multifractal atmospheric dynamics tested in the tropics: part I, horizontal scaling and self criticality, *Nonlinear Proc. Geophys.*, **1**, 105–114.

Cox, R. A. and Hayman, G. (1988), The stability and chemistry of dimers of the ClO radical: implications for Antarctic ozone, *Nature,* **332**, 796–800.

Danielsen, E. F. (1964), *Report on Project Springfield,* DASA 1517, Defense Atomic Support Agency, Washington, D. C.

Danielsen, E. F. (1968), Stratospheric-tropospheric exchange based upon radioactivity, ozone and potential vorticity, *J. Atmos. Sci.,* **25**, 502–518.

Davidson, P. A. (2004), *Turbulence,* Oxford University Press.

Davies, T., Cullen, M. J. P., Malcolm, A. J., Mawson, M. H., Staniforth, A., White, A. A., and Wood, N. (2005), A new dynamical core for the Met Office's global and regional modelling of the atmosphere, *Q. J. R. Meteorol. Soc.,* **131**, 1759–1782.

Dewar, R. (2003), Information theory explanation of the fluctuation theorem, maximum entropy production and self-organized criticality in non-equilibrium stationary states, *J. Phys. A: Math. Gen.,* **36,** 631–641.

Dewar, R. C. (2005a), Maximum entropy production and the fluctuation theorem, *J. Phys. A: Math. Gen.,* **38,** L371–L381.

Dewar, R. C. (2005b), Maximum entropy production and non-equilibrium statistical mechanics, Chapter 4 in *Non-equilibrium Thermodynamics and the Production of Entropy*, A. Kleidon and R. D. Lorenz, eds., pp. 41–55, Springer, Berlin.

Dobson, G. M. B., Brewer, A. W. and Cwilong, B. M. (1945), Meteorology of the lower stratosphere, *Proc. Roy. Soc. A*, 185, 144–175.

Dorfman, J. R. and Cohen, E. G. D. (1970), Velocity correlation functions in two and three dimensions, *Phys. Rev. Lett.*, **25**, 1257–1260.

Dougherty, J. P. (1994), Foundations of non-equilibrium statistical mechanics, *Phil. Trans. R. Soc. Lond. A* **346,** 259–305.

Dutton, J. A. (1986), *The Ceaseless Wind*, Dover, New York.

Eady, E. T. (1950), *The Cause of the General Circulation of the Atmosphere*, in Centenary Proceedings of the Royal Meteorological Society, pp. 156–172, Royal Meteorological Society, London.

Eady, E. T. (1951), Discussion remark, *Q. J. R. Meteorol. Soc.,* **77,** 316. See Batchelor (1951).

Eady, E. T. and Sawyer, J. S. (1951), Dynamics of flow patterns in extratropical regions, *Q. J. R. Meteorol. Soc.,* **77,** 531–551.

Edouard, S., Legras, B., Lefèvre, F., and Eymard, R. (1996), The effect of mixing on ozone depletion in the Arctic, *Nature,* **384,** 444–447.

Einstein, A. (1905), Die von Molekularkinetischen Theorie von Wärme geforderte Bewegung von in ruhende Flüssigkeiten suspendierten Teilchen, *Ann. Phys.,* **17,** 549–560.

Ertel, H. (1942) Ein neuer hydrodynamischer Wirbelsatz, *Meteorol. Z.*, **59,** 277–281, 385.

Evans, D. J. and Searles, D. J. (2002), The fluctuation theorem, *Adv. Phys.,* **51,** 1529–1585.

Fahey, D. W., Kelly, K. K., Ferry, G. V., Poole, L. R., Wilson, J. C., Murphy, D. M., Loewenstein, M., and Chan, K. R. (1989), In situ measurements of total reactive nitrogen, total water, and aerosol in a polar stratospheric cloud in the Antarctic, *J. Geophys. Res.,* **94,** 11,299–11,315.

Fahey, D. W., Gao, R.-S., Carslaw, K. S., Kettleborough, J., Popp, P. J., Northway, M. J., Holecek, J. C., Ciciora, S. C., McLaughlin, R. J., Thompson, T. L., et al. (2001), The detection of large HNO_3-containing particles in the winter Arctic stratosphere, *Science,* **291**, 1026–1031.

Falkovich, G., Fouxon, A., and Stepanov, M. G. (2002), Acceleration of rain initiation by cloud turbulence, *Nature,* **419,** 151–154.

Falkovich, G., Stepanov, M. G., and Vucelja, M. (2006), Rain initiation time in turbulent warm clouds, *J. Appl. Meteorol. Clim.,* **45,** 591–599.

Farman, J. C., Murgatroyd, R. J., Silnickas, A. M., and Thrush, B. A. (1985), Ozone photochemistry in the Antarctic stratosphere in summer, *Q. J. R. Meteorol. Soc.,* **111,** 1013–1028.

Feely, H. W., and Spar, J. (1960), Tungsten—185 from nuclear bomb tests as a tracer for stratospheric meteorology, *Nature,* **188**, 1062–1064.

Finlayson-Pitts, B. J. and Pitts, J. N. Jr. (2000), *Chemistry of the Upper and Lower Atmosphere*, Academic Press, San Diego.

Foot, J. S. (1984), Aircraft measurements of the humidity in the lower stratosphere from 1977 to 1980 between 45°N and 65°N, *Q. J. R. Meteorol. Soc.,* **110**, 303–319.

Frisch, U. (1995), *Turbulence: The Legacy of A.N. Kolmogorov*, Cambridge University Press.

Galewsky, J., Sobel, A., and Held, I. (2005), Diagnosis of subtropical humidity dynamics using tracers of last saturation, *J. Atmos. Sci.,* **62,** 3353–3367.

Gallavotti, G. (1999), *Statistical Mechanics*, Springer, Berlin.

Gao, R.-S., Richard, E. C., Popp, P. J., Toon, G. C., Hurst, D. F., Newman, P. A., Holecek, J. C., Northway, M. J., Fahey, D. W., Danilin, M. Y., Sen, B., Aikin, K., Romashkin, P. A., Elkins, J. W., Webster, C. R., Schauffler, S. M., Greenblatt, J. B., McElroy, C. T., Lait, L. R., Bui, T. P., and Baumgardner, D. (2001) Observational evidence for the role of denitrification in Arctic stratospheric ozone loss, *Geophys. Res. Lett.,* **28,** 2879–2882.

Gao, Y. Q. and Marcus, R. A. (2001), Strange and unconventional isotope effects in ozone formation, *Science,* **293,** 259–263.

Gary, B. L. (1989), Observational results using the Microwave Temperature Profiler during the Airborne Antarctic Ozone Experiment, *J. Geophys. Res.* **94,** 11,223–11,231.

Gary, B. L. (2006), Mesoscale temperature fluctuations in the stratosphere, *Atmos. Chem. Phys.* **6,** 4577–4589.

Gauss, M., Myhre, G., Isaksen, I. S. A., Grewe, V., Pitari, G., Wild, O., Collins, W. J., Dentener, F. J., Ellingsen, K., Gohar, L. K., Hauglustaine, D. A., Iachetti, D., Lamarque, J. F., Mancini, E., Mickley, L. J., Prather, M. J., Pyle, J. A., Sanderson, M. G., Shine, K. P., Stevenson, D. S., Sudo, K., Szopa, S., and Zeng, G. (2006), Radiative forcing since pre-industrial times due to ozone change in the troposphere and the lower stratosphere, *Atmos. Chem. Phys.,* **6,** 575–599.

Goody, R. M. and Yung, Y. L. (1989), *Atmospheric Radiation*, 2nd Ed., Oxford University Press, Oxford.

Grad, H. (1958), Principles of the Kinetic Theory of Gases, in *Handbuch der Physik*, Band **12,** S. Flugge, ed., pp. 205–294, Springer-Verlag, Berlin.

Grad, H. (1983), Singular and Non-uniform limits of Solutions of the Boltzmann Equations, in *Rarefied Gas Dynamics*, K. Karamcheti, ed., pp. 37–53, Academic Press, New York.

Graedel, T. E., Hawkins, D. T., and Claxton, L. D. (1986), *Atmospheric Chemical Compounds,* Academic Press, Orlando, Florida.

Harries, J. E. (1997), Atmospheric radiation and atmospheric humidity, *Q. J. R. Meteorol. Soc.,* **123,** 2173–2186.

Harris, N. R. P., Ancellet, G., Bishop, L., Hofmann, D. J., Kerr, J. B., McPeters, R. D., Prendez, M., Randel, W. J., Staehelin, J., Subbaraya, B. H., Volz-Thomas, A., Zawodny, J., and Zerefos, C. S. (1997), Trends in stratospheric and free tropospheric ozone, *J. Geophys. Res.,* **102,** 1571–1590.

Harris, S. (1971) *An Introduction to the Theory of the Boltzmann Equation,* Holt, Rinehart and Winston, New York (re-issued by Dover Books, 2004).

Hayman, G. D., Davies, J. M., and Cox, R. A. (1986), Kinetics of the reaction ClO + ClO → products and its potential relevance to Antarctic ozone, *Geophys. Res. Lett.,* **13,** 1347–1350.

Heisenberg, W. (1948), Zur statistichen Theorie der Turbulenz, *Zeit. f. Phys.,* **124,** 628–657.

Helliwell, N. C., Mackenzie, J. R., and Kerley, M. J. (1957), Some further observations from aircraft of frost point and temperature up to 50,000 ft., *Q. J. R. Meteorol. Soc.,* **83,** 257–262.

Hinshelwood, C. N. (1940), *The Kinetics of Chemical Change*, Oxford University Press, Oxford.

Hinshelwood, C. N. (1951), *The Structure of Physical Chemistry*, Oxford University Press, Oxford.

Hirschfelder, J. O., Curtiss, C. F., and Bird, R. B. (1964), *Molecular Theory of Gases and Liquids,* 2nd edition, Wiley, New York.

Hock, T. F. and Franklin, J. L. (1999), The NCAR GPS dropwindsonde, *Bull. Amer. Meteorol. Soc.,* **80,** 406–420.

Hoppel, K., Bevilacqua, R. M., Nedoluha, G., Deniel, C., Lefèvre, F., Lumpe, J. D., Fromm, M. D., Randall, C. E., Rosenfield, J. E., and Rex, M. (2002), POAM III observations of Arctic ozone loss for the 1999/2000 winter, *J. Geophys. Res.,* **107(D20),** Art. No. 8262, doi: 10.1029/2001JD000476.

Hoskins, B. J., McIntyre, M. E., and Robertson, A. W. (1985). On the use and significance of isentropic potential vorticity maps, *Q. J. R. Meteorol. Soc.*, **111,** 877–946.

Hovde, S. J., Tuck, A. F., Lovejoy, S., and Schertzer, D. (2007a), Vertical scaling of the atmosphere. I. Dropsondes from 13 km to the surface, *Q. J. R. Meteorol. Soc.,* to be submitted.

Hovde, S. J., Tuck, A. F., Lovejoy, S., and Schertzer, D. (2007b), Vertical scaling of the atmosphere. II. Balloons from 40 km to the surface, *Q. J. R. Meteorol. Soc.*, to be submitted.

IPCC Third Assessment Report (2001), *Climate Change 2001: The Scientific Basis,* Chapter 8, Cambridge University Press.

Jaynes, E.T. (1957a), Information theory and statistical mechanics, *Phys. Rev.* **106,** 620–630.

Jaynes, E. T. (1957b), Information theory and statistical mechanics II, *Phys. Rev.* **108,** 171–190.

Jaynes, E. T. (1965), Gibbs vs. Boltzmann entropies, *Amer. J. Phys.*, **33,** 391–398.

Jaynes, E. T. (2003), Bretthurst, G. L., ed., *Probability Theory: The Logic of Science,* Cambridge University Press, Cambridge.

Johnston, H. S. (1971), Reduction of stratospheric ozone by nitrogen oxide catalysts from supersonic transport exhaust, *Science,* **173,** 517–522.

Jones, R. L., Austin, J., McKenna, D. S., Anderson, J. G., Fahey, D. W., Farmer, C. B., Heidt, L. E., Kelly, K. K., Murphy, D. M., Proffitt, M. H., Tuck, A. F., and Veddder, J. F. (1989), Lagrangian photochemical modeling studies of the 1987 Antarctic spring vortex, 1, Comparison with AAOE observations, *J. Geophys. Res.,* **94,** 11,529–11,558.

Kadau, K., Germann, T. C., and Lomdahl, P. S. (2004), Large-scale molecular-dynamics simulation of 19 billion particles, *Int. J. Mod. Phys.,* **15,** 193–201.

Kelly, K. K., Tuck, A. F., Murphy, D. M., Proffitt, M. H., Fahey, D. W., Jones, R. L., McKenna, D. S., Loewenstein, M., Podolske, J. R., Strahan, S. E., Ferry, G. V., Chan, K. R., Vedder, J. F., Gregory, G. L., Hypes, W. E., McCormick, M. P., Browell, E. V., and Heidt, L. E. (1989), Dehydration in the lower Antarctic stratosphere during late winter and early spring 1987, *J. Geophys. Res.,* **94,** 11,317–11,358.

Kelly, K. K., Tuck, A. F., Heidt, L. E., Loewenstein, M., Podolske, J. R., Strahan, S. E., and Vedder, J. F. (1990), A comparison of ER-2 measurements of stratospheric water vapor between the 1987 Antarctic and 1989 Arctic airborne missions, *Geophys. Res. Lett.,* **17,** 465–468.

Kelly, K. K., Tuck, A. F., and Davies, T. (1991), Wintertime asymmetry of upper tropospheric water between the Northern and Southern Hemispheres, *Nature,* **353,** 244–247.

Kelly, K. K., Proffitt, M. H., Chan, K. R., Loewenstein, M., Podolske, J. R., Strahan, S. E., Wilson, J. C., and Kley, D. (1993), Water vapor and cloud water measurements over Darwin during the STEP tropical mission, *J. Geophys. Res.,* **98,** 8713–8723.

Kharchenko, V. and Dalgarno, A. (2004), Thermalization of fast $O(^1D)$ atoms in the stratosphere and mesosphere, *J. Geophys. Res.,* **109,** Art. No. D18311, doi: 10.1029/2004JD004597.

Kleidon, A. and Lorenz, R. D. (2005), Entropy production by Earth system processes, Chapter 1 in *Non-equilibrium Thermodynamics and the Production of Entropy,* pp. 1–20, Springer, Berlin.

Kolmogorov, A. N. (1962), A refinement of previous hypotheses concerning the local structure of turbulence in a viscous incompressible fluid at high Reynolds number, *J. Fluid. Mech.* **13,** 82–85.

Koscielny-Bunde, E., Bunde, A., Havlin, S., Roman, H. E., Goldreich, Y., and Schellnhuber, H-J. (1998), Indication of a universal persistence law governing atmospheric variability, *Phys. Rev. Lett.,* **81,** 729–732.

Kolmogorov, A. N. (1991), *Proc. R. Soc. Lond.* A **434**, 9–17 [English translations of 1941 Russian papers].

Landau, L. D. and Lifshitz, E. M. (2003) *Fluid Mechanics*, 2nd edition, *Course of Theoretical Physics*, Vol. 6, Chapter 3, Butterworth Heinemann, Oxford.

Lazarev, A., Schertzer, D., Lovejoy, S., and Chigirinskaya, Y. (1994), Unified multifractal atmospheric dynamics tested in the tropics: part II, vertical scaling and generalized scale invariance, *Nonlinear Process. Geophys.,* **1,** 115–123.

Loewenstein, M., Podolske, J. R., Chan, K. R., and Strahan, S. E. (1989), Nitrous oxide as a dynamical tracer in the 1987 Airborne Antarctic Ozone Experiment, *J. Geophys. Res.,* **94,** 11,589–11,598.

Lovejoy, S., Schertzer, D., and Stanway, J. D. (2001), Direct evidence of multifractal cascades from planetary scales down to 1 km, *Phys. Rev. Lett.*, **86**, 5200–5203.

Lovejoy, S., Schertzer, D., and Tuck, A. F. (2004), Fractal aircraft trajectories and nonclassical turbulent statistics, *Phys. Rev. E*, **70**, Art. No. 036306, doi: 10.1103/PhysRevE.70.036306.

Lovejoy, S., Tuck, A. F., Hovde, S. J., and Schertzer, D. (2007a), Is isotropic turbulence relevant in the atmosphere?, *Geophys. Res. Lett.,* **34,** doi:10.1029/2007GL029359.

Lovejoy, S., Tuck, A. F., Hovde, S. J., and Schertzer, D. (2007b), Do stable atmospheric layers exist?, *Geophys. Res. Lett.,* **34,** submitted.

MacDonald, A. E. (2005), A global profiling system for improved weather and climate prediction, *Bull. Amer. Meteorol. Soc.,* **86,** 1747–1764.

Mandelbrot, B. B. (1974), Intermittent turbulence in self-similar cascades: divergence of high moments and dimension of the carrier, *J. Fluid Mech.*, **62**, 331–358.

Mandelbrot, B. B. (1983), *The Fractal Geometry of Nature,* W. H. Freeman, New York.

Mandelbrot, B. B. (1998), *Multifractals and 1/f Noise*, Springer, Berlin.

Marenco, A., Youget, H., Nédélec, P., Pagés, G.-P., and Karcher, F. (1994), Evidence of a long-term increase in tropospheric ozone from Pic du Midi data series: consequences: positive radiative forcing, *J. Geophys. Res.*, **99**, 16, 617–16, 632.

Marenco, A., Thouret, V., Nédélec, P., Smit, H., Helten, M., Kley, D., Karcher, F., Simon, P., Law, K., Pyle, J., Pöschmann, G., Von Wrede, R., Hume, C., and Cook, T. (1998), Measurement of ozone and water vapor by Airbus in-service aircraft: the MOZAIC program, an overview, *J. Geophys. Res.,* **103,** 25, 631–25, 642.

Mauersberger, K. (1981), Measurement of heavy ozone in the stratosphere, *Geophys. Res. Lett.,* **8,** 935–937.

Mauersberger, K., Erbacher, B., Krankowsky, D., Gunther, J., and Nickel, R. (1999), Ozone isotope enrichment: isotopomer-specific rate coefficients, *Science,* **283,** 370–372.

Mauersberger, K., Krankowsky, D., Janssen, C., and Schinke, R. (2005), Assessment of the ozone isotope effect, *Adv. Atom. Mol. Opt. Phys.,* **50,** 1–54.

McElroy, C. T. (1995), A spectroradiometer for the measurement of direct and scattered solar irradiance from on-board the NASA ER-2 high altitude research aircraft, *Geophys. Res. Lett.*, **22**, 1361–1364.

McElroy, M. B., Salawitch, R. J., Wofsy, S. C., and Logan, J. A. (1986), Antarctic ozone: reductions due to synergistic interactions of chlorine and bromine, *Nature,* **321,** 759–762.

Mitchell, J. F. B. (2004), Can we believe predictions of climate change? *Q. J. R. Meteorol. Soc.,* **130,** 2341–2360.

Molina, M. J. and Rowland, F. S. (1974), Stratospheric sink for chlorofluoromethanes: chlorine atom catalyzed destruction of ozone, *Nature,* **249,** 810–814.

Molina, M. J. and Molina, L. T. (1987), Production of Cl_2O_2 from the self-reaction of the ClO radical, *J. Phys. Chem.*, **91**, 433–436, 1987.

Murgatroyd, R. J. (1957), Winds and temperatures between 20 km and 100 km — a review, *Q. J. R. Meteorol. Soc.*, **83**, 417–458.

Murgatroyd, R. J. (1965), Ozone and water vapour in the upper troposphere and lower stratosphere, in *Meteorological Aspects of Atmospheric Radioactivity*, W. M. O. No. 169, pp. 68–94, World Meteorological Organization, Geneva.

Murgatroyd, R. J. and Clews, C. J. B. (1949), *Wind at 100,000 ft. over South-East England*, Meteorological Office Geophysical Memoirs No. 83, HMSO, London.

Murgatroyd, R. J., Goldsmith, P., and Hollings, W. E. H. (1955), Some recent measurements of humidity from aircraft up to heights of about 50,000 ft. over southern England, *Q. J. R. Meteorol. Soc.*, **81**, 533–537.

Murgatroyd, R. J. and Singleton, F. (1961), Possible meridional circulations in the stratosphere and mesosphere, *Q. J. R. Meteorol. Soc.*, **87**, 125–135.

Murphy, D. M. and Gary, B. L. (1995), Mesoscale temperature fluctuations and polar stratospheric clouds, *J. Atmos. Sci.*, **52**, 1753–1760.

Murphy, D. M. (2003), Dehydration in cold clouds is enhanced by a transition from cubic to hexagonal ice, *Geophys. Res. Lett.*, **30**, Art. No. 2230, doi:10.1029/2003GL018566.

Nastrom, G. D. and Gage, K. S. (1985), A climatology of atmospheric wave number spectra of wind and temperature observed by commercial aircraft, *J. Atmos. Sci.*, **42**, 950–960.

Newman, P. A., Fahey, D. W., Brune, W. H., and Kurylo, M. J. (1999), Preface, *J. Geophys. Res.*, **104**, 26, 481–26, 495.

Newman, P. A., Harris, N. R. P., Adriani, A., Amanatidis, G. T., Anderson, J. G., Braathen, G. O., Brune, W. H., Carslaw, K. S., Craig, M. S., DeCola, P. L., Guirlet, M., Hipskind, R. S., Kurylo, M. J., Küllmann, H., Larsen, N., Mégie, G. J., Pommereau, J.-P., Poole, L. R., Schoeberl, M. R., Stroh, F., Toon, O. B., Trepte, C. R., and Van Roozendael, M. (2002), An overview of the SOLVE/THESEO 2000 campaign, *J. Geophys. Res.*, **107(D20)**, Art. No. 8259, doi: 10.1029/2001JD001303.

Onsager, L. (1945), The distribution of energy in turbulence, *Phys. Rev.*, **68**, 285.

Onsager, L. (1949), Statistical hydrodynamics, *Nuovo Cimento*, **6**, 279–287.

Palmer, T. N. (2001), A nonlinear perspective on model error: a proposal for non-local stochastic-dynamical parameterization in weather and climate prediction models, *Q. J. R. Meteorol. Soc.*, **127**, 279–304.

Paltridge, G. W. (1975), Global dynamics and climate-a system of minimum entropy exchange, *Q. J. R. Meteorol. Soc.*, **101**, 475–484.

Paltridge, G. W. (2001), A physical basis for a maximum of thermodynamic dissipation of the climate system, *Q. J. R. Meteorol. Soc.*, **127**, 305–313.

Paltridge, G. W. (2005), Stumbling into the MEP racket: an historical perspective, Chapter 3 in *Non-equilibrium Thermodynamics and the Production of Entropy*, pp. 33–40, Springer, Berlin.

Parisi, G. and Frisch, U. (1985), On the singularity structure of fully developed turbulence, in *Turbulence and Predictability in Geophysical Fluid Dynamics*, Proc. Int. School Phys. 'E. Fermi', 1983, Varenna, Italy, pp 84–87, North Holland, Amsterdam.

Pedlosky, J. (1987), *Geophysical Fluid Dynamics*, 2nd edition Springer-Verlag, New York.

Plumb, R. A., Waugh, D. W., and Chipperfield, M. P. (2000), The effects of mixing on tracer relationships in the polar vortices, *J. Geophys. Res.*, **105**, 10,047–10,062.

Proffitt, M. H. and McLaughlin, R. J. (1983), Fast-response dual-beam UV-absorption ozone photometer suitable for use on stratospheric balloons, *Rev. Sci. Instrum.*, **54**, 1719–1728.

Proffitt, M. H., Fahey, D. W., Kelly, K. K., and Tuck, A. F. (1989), High-latitude ozone loss outside the Antarctic ozone hole, *Nature*, **342**, 233–237.

Proffitt, M. H., Aikin, K., Tuck, A. F., Margitan, J. J., Webster, C. R., Toon, G. C., and Elkins, J. W. (2003), Seasonally averaged ozone and nitrous oxide in the northern hemisphere lower stratosphere, *J. Geophys. Res.*, **108(D3)**, Art. No. 4110, doi:10.1029/2002JD002657.

Pyle, J. A., Braesicke, P., and Zeng, G. (2005), Dynamical variability in the modelling of chemistry-climate interactions, *Faraday Discuss.*, **130**, 27–39.

Rapaport, D. C. (2004), *The Art of Molecular Dynamics Simulation*, 2nd edition,Cambridge University Press.

Reed, R. J. and Danielsen, E. F. (1959), Fronts in the vicinity of the tropopause, *Arch. Meteorol. Geophys. Bioklim.*, **11**, 1–17.

Reed, R. J., and German, K. E. (1965), A contribution to the problem of stratospheric diffusion by large-scale mixing, *Mon. Wea. Rev.*, **93**, 313–321.

Reiter, E. R. (1963), *Jet Stream Meteorology*, University of Chicago Press, Chicago.

Richard, E. C., Aikin, K. C., Andrews, A. E., Daube, B. C. Jr., Gerbig, C., Wofsy, S. C., Romashkin, P. A., Hurst, D. F., Ray, E. A., Moore, F. L., Elkins, J. W., Deshler, T., and Toon, G. C. (2001), Severe chemical ozone loss inside the Arctic polar vortex during winter 1999–2000 inferred from *in situ* airborne measurements, *Geophys. Res. Lett.*, **28**, 2197–2200.

Richard, E. C., Tuck, A. F., Aikin, K. C., Kelly, K. K., Herman, R. L., Troy, R. F., Hovde, S. J., Rosenlof, K. H., Thompson, T. L., and Ray, E. A. (2006), High-resolution airborne profiles of CH_4, O_3 and water vapor near tropical Central America in late January – early February 2004, *J. Geophys. Res.*, **111**, D13304, doi:10.1029/2005JD006513.

Richardson, L. F. (1926), Atmospheric diffusion shown on a distance-neighbour graph, *Proc. R. Soc. Lond.* A **110**, 709–737.

Riehl, H., Alaka, M. A., Jordan, C. L., and Renard, R. J. (1954), The Jet Stream, *Meteorological Monographs*, Volume 2, Number 7, American Meteorological Society, Boston, MA.

Rosenlof, K. H., Tuck, A. F., Kelly, K. K., Russell, J. M., and McCormick, M. P. (1997), Hemispheric asymmetries in water vapor and inferences about transport in the lower stratosphere, *J. Geophys. Res.*, **102**, 13, 213–13, 234.

Rossby, C-G. (1940), Planetary flow patterns in the atmosphere, *Q. J. R. Meteorol. Soc.*, **66**, Supplement, 68–87.

Rossby, C. G. (1947), On the distribution of angular velocity in gaseous envelopes under the influence of large scale horizontal mixing processes, *Bull. Amer. Meteorol. Soc.*, **28**, 53–68.

Russell, J. M., Tuck, A. F., Gordley, L. L., Park, J. H., Drayson, S. R., Harries, J. E., Cicerone, R. J., and Crutzen, P. J. (1993), HALOE Antarctic observations in the spring of 1991, *Geophys. Res. Lett.*, **20**, 719–722.

Samorodinsky, G. and Taqqu, M. S. (1994), *Stable Non-Gaussian Random Processes*, Chapman and Hall, New York.

Sawyer, J. S. (1951), The dynamical systems of the lower stratosphere, *Q. J. R. Meteorol. Soc.*, **77**, 480–483.

Schertzer, D. and Lovejoy, S. (1985), The dimension and intermittency of atmospheric dynamics, in *Turbulent Shear Flows*, Vol. **4**, pp. 7–33, Springer, New York, USA.

Schertzer, D. and Lovejoy, S. (1987), Physical modeling and analysis of rain and clouds by anisotropic scaling multiplicative processes, *J. Geophys. Res.*, **92**, 9693–9714.

Schertzer, D. and Lovejoy, S. (1991), eds., *Nonlinear Variability in Geophysics: Scaling and Fractals*, Chapters 3, 4, 7 and 8, Kluwer Academic Publishers, Dordrecht.

Schmitt, F., Schertzer, D., Lovejoy, S., and Brunet, Y. (1994), Empirical study of multifractal phase transitions in atmospheric turbulence, *Nonlinear Process. Geophys.*, **1**, 94–104.

Searle, K. D., Chipperfield, M. P., Bekki, S., and Pyle, J. A. (1998), The impact of spatial averaging on calculated polar ozone loss: 2. Theoretical analysis, *J. Geophys. Res.*, **103**, 25, 409–25, 416.

Seuront, L., Schmitt, F., Lagadeux, Y., Schertzer, D., and Lovejoy, S. (1999), Universal multifractal analysis as a tool to characterize multiscale intermittent patterns: example of phytoplankton distribution in turbulent coastal waters, *J. Plankton Res.*, **21**, 877–922.

Shapiro, M. A. (1980), Turbulent mixing within troposphere folds as a mechanism for the exchange of chemical constituents between the stratosphere and troposphere, *J. atmos. sci.*, **37**, 994–1004.

Shapiro, M. A., Reiter, E. R., Cadle, R. D., and Sedlacek, W. A. (1980), Vertical mass and trace constituent transports in the vicinity of jet streams, *Arch. Meteorol. Geophys. Bioklim.*, **B28**, 193–206.

Shraiman, B. I. and Siggia, E. D. (2000), Scalar turbulence, *Nature,* **405**, 639–646.

Smoluchowski, M. (1906), Kinetische Theorie der Brownsche Bewegung und der Suspensionen, *Ann. Phys.*, **21**, 756–780.

Sparling, L. C. (2000), Statistical perspectives on stratospheric transport, *Revs.Geophys.*, **38**, 417–436.

Succi, S. (2001), *The Lattice Boltzmann Equation for Fluid Dynamics and Beyond,* p. 134, Clarendon Press, Oxford.

Swartz, W. H., Lloyd, S. A., Kusterer, T. L., Anderson, D. E., McElroy, C. T., and Midwinter, C. (1999), A sensitivity study of photolysis rate coefficients during POLARIS, *J. Geophys. Res.*, **104**, 26, 725–26, 736.

Syroka, J. and Toumi, R. (2001), Scaling of central England temperature fluctuations?, *Atmos. Sci. Lett.*, **2**, 143–154.

Takahashi, K., Taniguchi, N., Sato, Y., and Matsumi, Y. (2002), Nonthermal steady state translational energy distributions of $O(^1D)$ atoms in the stratosphere, *J. Geophys. Res.*, **107(D16)**, Art. No. 4290, doi: 10.1029/2001JD001270.

Tan, D. G. H., Haynes, P. H., MacKenzie, A. R., and Pyle, J. A. (1998), Effects of fluid dynamical stirring and mixing on the deactivation of stratospheric chlorine, *J. Geophys. Res.*, **103**, 1585–1605.

Toohey, D. W., Anderson, J. G., Brune, W. H., and Chan, K. R. (1990), In situ measurements of BrO in the Arctic stratosphere, *Geophys. Res. Lett.*, **17**, 513–516.

Touchette, H. (2004), Temperature fluctuations and mixtures of equilibrium states in the canonical ensemble, Chapter 9 in *Nonextensive Entropy,* M. Gell-Mann and C. Tsallis, eds., pp. 159–176, Oxford University Press.

Toumi, R., Syroka J., Barnes C., and Lewis, P. (2001) Robust non-Gaussian statistics and long-range correlation of total ozone, *Atmos. Sci. Lett.*, **2**, 94–103.

Tripathi, O. P., Leblanc, T., McDermid, I. S., Lefèvre, F., Marchand, M., and Hauchecorne, A. (2006), Forecast, measurement, and modeling of an unprecedented polar ozone filament event over Mauna Loa Observatory, Hawaii, *J. Geophys. Res.*, **111**, Art. No. D020308.

Truesdell, C. (1952), Vorticity and the thermodynamic state in a gas flow, *Mem. Sci. Math.*, Fascicule **CXIX**, 1–53, Gauthier-Villars, Paris.

Truesdell, C. (1954), *The Kinematics of Vorticity,* Sections 78, 79, 98 and 104, University of Indiana Press, Bloomington.

Tsallis, C. (2004), Nonextensive statistical mechanics: construction and physical interpretation, Chapter 1, in *Nonextensive Entropy - Interdisciplinary Applications*, M. Gell-Mann and C. Tsallis, eds.,Oxford University Press.

Tsallis, C., Gell-Mann, M., and Sato, Y. (2005), Asymptotically scale-invariant occupancy of phase space makes the entropy S_q extensive, *Proc. Natl. Acad. Sci.,* **102**, 15,377–15,382.

Tuck, A. F. (1979), A comparison of one-, two- and three- dimensional model representations of stratospheric gases, *Phil. Trans. Roy. Soc.,* **A290**, 477–494.

Tuck, A. F. (1993), Use of ECMWF products in stratospheric measurement campaigns, Workshop Proceedings, *Stratosphere and Numerical Weather Prediction,* pp.73–105, European Centre for Medium-range Weather Forecasts, Reading.

Tuck, A. F. (1989), Synoptic and chemical evolution of the Antarctic vortex in late winter and early spring 1987, *J. Geophys. Res.*, **94,** 11, 687–11, 737.

Tuck, A. F., Watson, R. T., Condon, E. P., Margitan, J. J., and Toon, O. B. (1989), The planning and execution of ER-2 and DC-8 aircraft flights over Antarctica, August and September 1987, *J. Geophys. Res.,* **94**, 11, 181–11, 222.

Tuck, A. F., Davies, T., Hovde, S. J., Noguer-Alba, M., Fahey, D. W., Kawa, S. R., Kelly, K. K., Murphy, D. M., Proffitt, M. H., Margitan, J. J., Loewenstein, M., Podolske, J. R., Strahan, S. E., and Chan, K. R. (1992), Polar stratospheric cloud-processed air and potential vorticity in the northern hemisphere lower stratosphere at mid-latitudes during winter, *J. Geophys. Res.,* **97** , 7883–7904.

Tuck, A. F., Baumgardner, D., Chan, K. R., Dye, J. E., Elkins, J. W., Hovde, S. J., Kelly, K. K., Loewenstein, M., Margitan, J. J., May, R. D., Podolske, J. R., Proffitt, M. H., Rosenlof, K. H., Smith, W. L., Webster, C. R., and Wilson, J. C. (1997), The Brewer-Dobson circulation in the light of high altitude *in situ* aircraft observations, *Q. J. R. Meteorol. Soc.,* **123**, 1–69.

Tuck, A. F. and Proffitt, M. H. (1997), Comment on "On the magnitude of transport out of the Antarctic vortex", *J. Geophys. Res.,* **102**, 28, 215–28, 218.

Tuck, A. F., and Hovde, S. J. (1999), Fractal behavior of ozone, wind and temperature in the lower stratosphere, *Geophys. Res. Lett.,* **26**, 1271–1274.

Tuck, A. F., Hovde, S. J., and Proffitt, M. H. (1999), Persistence in ozone scaling under the Hurst exponent as an indicator of the relative rates of chemistry and fluid mechanical mixing in the stratosphere, *J. Phys. Chem. A,* **103**, 10, 445–10, 550.

Tuck, A. F., Hovde, S. J., Richard, E.C., Fahey, D.W., and Gao, R.-S. (2002), A scaling analysis of ER-2 data in the inner vortex during January–March 2000, *J. Geophys. Res.*, **108**(D5), Art. No. 8306, doi:10.1029/2001JD000879.

Tuck, A. F., Hovde, S. J., Gao, R.-S., and Richard, E. C. (2003a), Law of mass action in the Arctic lower stratospheric polar vortex January–March 2000: ClO scaling and the calculation of ozone loss rates in a turbulent fractal medium, *J. Geophys. Res.,***108(D15)**, Art. No. 4451, doi:10.1029/2002JD002832.

Tuck, A. F., Hovde, S. J., Kelly, K. K., Mahoney, M. J., Proffitt, M. H., Richard, E. C., and Thompson, T. L. (2003b), Exchange between the upper tropical troposphere and the lower stratosphere studied with aircraft observations, *J. Geophys. Res.*, **108(D23)**, Art. No. 4734, doi:10.1029/2003JD003399.

Tuck, A. F., Hovde, S. J., and Bui, T. P. (2004), Scale invariance in jet streams: ER-2 data around the lower-stratospheric polar night vortex, *Q. J. R. Meteorol. Soc.*, **130**, 2423–2444.

Tuck, A. F., Hovde, S. J., Richard, E. C., Gao, R.-S., Bui, T. P., Swartz, W. H., and Lloyd, S. A. (2005), Molecular velocity distributions and generalized scale invariance in the turbulent atmosphere, *Faraday Discuss.*, **130**, 181–193, doi:10.1039/b410551f.

Vaida, V., Solomon, S., Richard, E. C., Ruhl, E., and Jefferson, A. (1989), Photoisomerization of OClO - a possible mechanism for polar ozone depletion, *Nature,* **342**, 405–408.

Vaida, V., Daniel, J. S., Kjaergaard, H. G., Goss, L. M., and Tuck, A. F. (2001), Atmospheric absorption of near infrared and visible solar radiation by the hydrogen bonded water dimer, *Q. J. R. Meteorol. Soc.,* **127**, 1627–1643.

van Kampen, N. G. (2002), The road from molecules to Onsager, *J. Stat. Phys.,* **109**, 471–481.

Varotsos, C. (2005), Power law correlations in column ozone over Antarctica, *Int. J. Remote Sensing,* **26**, 3333–3342.

Vassilicos, J. C. (2002), Mixing in vortical, chaotic and turbulent flows, *Phil. Trans. R. Soc.,* **A360**, 2819–2837.

Vinnichenko, N. K., Pinus, N. Z., Shmeter, S. M. and Shur, G. N. (1980), *Turbulence in the Free Atmosphere,* 2nd edition, Plenum, New York.

Volz, A. and Kley, D. (1988), Evaluation of the Montsouris series of ozone measurements made in the 19th century, *Nature*, **332**, 240–242.

von Neumann, J. L., (1963) Recent Theories of Turbulence, in *Collected Works of John von Neumann*, Vol. **6**, pp. 437–472, Macmillan, NewYork.

von Weizsäcker, C. F., (1948), Das Spektrum der Turbulenz bei großen Reynoldschen Zahlen, *Zeit. f. Phys.*, **124**, 614–627.

Wagner, G., Birk, M., Gamache, R. R., and Hartmann, J.-M. (2005), Collisional parameters of H_2O lines: effect of temperature, *J. Quant. Spectr. Rad. Transfer,* **92**, 211–230.

Wang, G. M., Sevick, E. M., Mittag, E., Searles, D. J. and Evans, D. J. (2002), Experimental demonstration of violations of the second law of thermodynamics for small systems and short timescales, *Phys. Rev. Lett.,* **89**, Art. No. 050601.

Waters, J. W., Froidevaux, L., Read, W. G., Manney, G. L., Elson, L. S., Flower, D. A., Jarnot, R. F. and Harwood, R. S. (1993), Stratospheric ClO and ozone from the Microwave Limb Sounder on the Upper Atmosphere Research Satellite, *Nature,* **362**, 597–602.

Waterston, J. J. (1845, 1892), I. On the physics of media that are composed of free and perfectly elastic molecules in a state of motion. With an introduction by Lord Rayleigh, *Phil. Trans. R. Soc. London A,* **183**, 1–79.

Webster, C. R. and Heymsfield, A. J. (2003), Water isotope ratios D/H, O^{18}/O^{16}, O^{17}/O^{16} in and out of clouds map dehydration pathways, *Science,* **302**, 1742–1745.

White, A. A. (2002), A view of the equations of meteorological dynamics and various approximations, in *Large-Scale Atmosphere-Ocean Dynamics I*, J. Norbury and I.Roulstone(eds), pp. 19–27, Cambridge University Press.

White, A. A., Hoskins, B. J., Roulstone, I., and Staniforth, A. (2005), Consistent approximate models of the global atmosphere: shallow, deep, hydrostatic, quasi-hydrostatic and non-hydrostatic, *Q. J. R. Meteorol. Soc.,* **131**, 2081–2107.

Wilson, J. C., Loewenstein, M., Fahey, D. W., Gary, B., Smith, S. D., Kelly, K. K., Ferry, G. V., and Chan, K. R. (1989), Observations of condensation nuclei in the Antarctic Airborne Ozone Experiment: implications for new particle formation and polar stratospheric cloud formation, *J. Geophys. Res.,* **94**, 16,437–16,448.

Wofsy, S. C., Cohen, R. C., and Schmeltekopf, A. L. (1994), Overview: the Stratospheric Photochemistry, Aerosols and Dynamics expedition (SPADE) and Airborne Arctic Stratospheric Expedition II (AASE II), *Geophys. Res. Lett.,* **23**, 2535–2358.

Wonhas, A. and Vassilicos, J. C. (2002), Diffusivity dependence of ozone depletion over the mid-northern latitudes, *Phys. Rev. E,* **65**, Art. No. 051111.

Yang, H., and R. Pierrehumbert (1994), Production of dry air by isentropic mixing, *J. Atmos. Sci.,* **51**, 3437–3451.

Zwanzig, R. (2001), *Nonequilibrium Statistical Mechanics*, Oxford University Press, Oxford.

Bibliography

My reading was particularly informed by the following books, which are personal choices from a large literature. Important or relevant references from scientific journals have been cited in the body of the book. Somewhat arbitrarily, the books have been assigned to seven categories.

[i] Molecules. Two volumes which qualify for the adjective 'literary' are Hinshelwood's books from over half a century ago, *The Kinetics of Chemical Change* and *The Structure of Physical Chemistry*. Essential equations are fitted elegantly into lucid text; as expositions of the basics I found them still to be unrivalled. Four other books were useful: Benson's *The Foundation of Chemical Kinetics*, the third edition of Chapman and Cowlings' *The Mathematical Theory of Non-Uniform Gases*, the second edition of Hirschfelder, Curtiss, and Birds' *The Molecular Theory of Gases and Liquids* and the third edition of Liboff's *Kinetic Theory*. These books collectively contain a large body of derivations and formulae. The eighth edition of Atkins' *Physical Chemistry* provides a modern, textbook view embodying more recent developments. A thorough yet clear exposition may be found in the second editions of three volumes by Berry, Rice, and Ross, *The Structure of Matter, Matter in Equilibrium* and *Physical and Chemical Kinetics*. Chapters 20, 27, and 28 are particularly apposite. Two fine texts on atmospheric chemistry are the third edition of Wayne's *Chemistry of Planetary Atmospheres* and *Chemistry of the Upper and Lower Atmosphere* by Finlayson-Pitts and Pitts. Finally, Rapaport's *The Art of Molecular Dynamics Simulation* is a unique text, which would make a good starting point for anyone willing to tackle the molecular scale generation of atmospheric vorticity.

[ii] Turbulence. During the last decade or so, two books on this subject have appeared, Frisch's *Turbulence* and more recently Davidson's *Turbulence*. They are essential reading; I found the latter to be particularly enlightening. The second edition of *Turbulence in the Free Atmosphere* by Vinnichenko, Pinus, Shmeter, and Shur gives an account of the standard approach to its subject matter, including an interesting discussion in Chapter 2 of how an aircraft can be used to detect turbulence over a certain range of scales.

[iii] Fractals. This subject has grown into a vast literature, but I have restricted myself to Mandelbrot's *Multifractals and 1/f Noise*, which

reprises over 30 years of the author's publications, and *Non-Linear Variability in Geophysics*, edited and partly written by Schertzer and Lovejoy, upon which I have leaned heavily.

[iv] Fluid Mechanics. In the non-meteorological category, Batchelor's *An Introduction to Fluid Dynamics* and the second edition of Landau and Lifshitzs' *Fluid Mechanics*, Volume 6 of *Course of Theoretical Physics* were helpful. For dynamical meteorology I resorted most frequently to the second edition of Dutton's *The Ceaseless Wind* and to Gill's *Atmosphere – Ocean Dynamics*. Added note: When this manuscript was complete, G. K. Vallis's *Atmospheric and Oceanic Fluid Dynamics: Fundamentals and Large-Scale Circulation* (2006), Cambridge University Press, became available. It is an excellent modern account, thorough and pedagogical. Chapters 8–10 are of special interest in our context, giving a detailed treatment of the standard approach through 2D and 3D turbulence. Note however that there is no consideration of the multifractal approach, its intermediate dimensionality and the long-tailed PDFs which are a central part of the approach taken here.

[v] Meteorology. For atmospheric physics, Houghton's *The Physics of Atmospheres* (2002), Cambridge University Press, third edition and Salby's *Fundamentals of Atmospheric Physics* (1996), Academic Press, were clear texts. For basic applications of meteorology, J. F. R. McIlveen, *Fundamentals of Weather and Climate* (1992), Chapman and Hall, is a good source. Although out of print, F. H. Ludlam, *Clouds and Storms: The Behavior and Effect of Water in the Atmosphere* (1980), The Pennsylvania State University Press, is a book offering a unique and deep view of its subject matter.

[vi] Statistical Mechanics. The classical texts on equilibrium statistical mechanics combine depth with breadth: Tolman's *The Principles of Statistical Mechanics* and Fowler's *Statistical Mechanics*. The third edition of Part 1 of Landau and Lifshitzs' *Statistical Physics*, Volume 5 in their *Course of Theoretical Physics* is particularly penetrating and well written. There are some recent books on non-equilibrium statistical mechanics which I found to be very useful. Zwanzig's *Non-Equilibrium Statistical Mechanics* is a clear exposition. Balescu's *Statistical Dynamics* considers a wide range of material in mathematical detail. Chen's recent *A Non-Equilibrium Statistical Mechanics Without the Assumption of Molecular Chaos* is an herculean effort, with some insightful text embedded in the mathematics. To work through the latter would probably be as demanding an exercise as it must have been to write it, even for professional mathematicians. *Non-equilibrium Thermodynamics and the Production of Entropy*, edited by Kleidon and Lorenz, is a collection of chapters covering the production of entropy and its maximization in a wide range of macroscopic systems. The

chapters by Dewar and by Paltridge are particularly illuminating as regards principle in our context.

[vii] Probability. I have used two books from a large field, Volume 2 of *An Introduction to Probability Theory and its Applications* by Feller and *Stable Non-Gaussian Random Processes* by Samorodinsky and Taqqu.

Glossary

This book unavoidably uses terminology specific to both physical chemistry and meteorology; a glossary is therefore provided. Should some terms used still be unfamiliar or require further explanation, the meteorological reader seeking physico-chemical enlightenment should first try the books by Hinshelwood and by Berry, Rice, and Ross that are listed in the bibliography. For physical chemists seeking help with dynamical meteorology, the books there by Dutton and by Gill are recommended. The books by Houghton and by Salby are recommended for atmospheric physics, and McIlveen's book is excellent for the basic principles of weather and climate. There are two comprehensive meteorological glossaries: *Meteorological Glossary*, The Stationery Office Books (1991) and *Glossary of Meteorology*, second edition, American Meteorological Society (2000).

AAOE. Airborne Antarctic Ozone Experiment. A mission based in Punta Arenas, on the Straits of Magellan, during August and September 1987, using the NASA ER-2 and DC-8 aircraft to investigate the ozone hole. See *J. Geophys. Res.*, **94**, 11,179–11,737 and 16,437–16,857 (1989).

AASE. Airborne Arctic Stratospheric Expedition. Following AAOE, it was realized that the ozone loss occurring in the outer regions of the Antarctic vortex did so in conditions that could be found in the interior of the Arctic vortex. AASE was essentially an Arctic repeat of AAOE, based at Stavanger in southern Norway from December 1988 to February 1989. See *Geophys. Res. Lett.*, **17**, 313–564 (1990). AASE-II was based in Bangor, Maine and was designed to follow the evolution of the Arctic vortex and its midlatitude environs from autumn 1991 to spring 1992. See *Geophys. Res. Lett.*, **20**, 2415–2578 (1993) and *Science*, **261**, 1128–1158 (1993).

Absolute vorticity. Defined in Eq. 3.10. as the curl of the velocity vector. In the atmosphere, the relative vorticity is taken as the component along the gravitationally aligned, that is vertical, axis of spin of a local air element. It is twice the angular velocity. The absolute vorticity is the sum of the relative vorticity plus the vorticity arising from the planetary rotation, see Eq. 3.17. The planetary vorticity is expressed as $f = 2\boldsymbol{\Omega} \sin\phi$, the Coriolis parameter, where $\boldsymbol{\Omega}$ is the planetary angular velocity and ϕ is latitude. As a result, the polar regions have high absolute vorticity while the tropics have low values.

ACCENT. (Atmospheric Chemistry of Combustion Emissions Near the Tropopause). An airborne mission using the NASA WB57F from Ellington Field, Houston, Texas, in spring and autumn 1999. See, for example, *J. Geophys. Res.*, **109**, D05310, doi: 10.1029/2003JD003942 (2004).

Adiabatic. A condition in which there is no loss or gain of energy by the entity being considered. An air mass obeying this condition conserves potential temperature; it is isentropic.

Advection. If a point on the Earth's surface is considered, the local rate of change of an atmospheric quantity (say temperature, T) is given by $\partial T/\partial t$, which is the sum of two quantities, the total or material derivative in the air mass being blown over the point, dT/dt, plus the rate of change arising from the fact that different air masses are being blown over the point, $-\boldsymbol{u}\cdot\nabla T$. The latter term is the advection; it is, for example, a description of the fact that a wind from the north during winter at a point in boreal midlatitudes feels cold.

Anticyclone. Anticyclonic flow has relative vorticity characteristic of rotation in the opposite sense to that of the Earth. In the troposphere such systems are generally of order a thousand kilometres in horizontal scale, have high pressure at the surface, and have descending motion in the troposphere, with convergence in the upper troposphere and divergence in the lower troposphere. The airflow is clockwise in the Northern Hemisphere and anticlockwise in the Southern Hemisphere.

Antipersistence. A condition occurring in the two-point correlation analysis of a one-dimensional data series when, for all choices of sampling interval, neighbouring intervals have opposite correlation. In the absence of intermittency, it corresponds to a Hurst exponent of zero. Unlike persistent behaviour, antipersistence is comparatively rare in nature.

ASHOE/MAESA. (Airborne Southern Hemisphere Ozone Experiment/ Measurements for Assessing the Effects of Stratospheric Aircraft). A mission using the NASA ER-2 from Christchurch, New Zealand, between March and October 1994. The objective was to examine the relation of the vortex and its ozone hole to midlatitudes over its evolution from autumn to spring. See *J. Geophys. Res.*, **102**, 3899–3949 and 13,113–13,299 (1997).

Autocorrelation. Autocorrelation in a time series is a measure of how well the series matches a time-shifted version of itself, as a function of the magnitude of the time shift. See also pp. 456–460 in Davidson (2004). It is useful for revealing phenomena that depend on the temporal ordering in the series; for example, periodic frequencies or long-range correlations.

Baroclinic instability. A condition in which the pressure surfaces intersect the temperature surfaces, most importantly in midlatitudes, in such a way that wave-like motions with wavelengths of thousands of kilometres amplify in intensity. It is the basic mechanism of cyclogenesis, the generation of low-pressure cyclonic circulations, and is important in weather front generation and energy exchange.

Biogeochemical. The adjective describing the flux of elements through molecular states involved in living, geological and chemical systems in the atmosphere, ocean, land, and biosphere. Many such transformations are involved in the carbon, nitrogen, oxygen, sulphur, and phosphorus cycles, for example.

Boltzmann H-equation. This is a nonlinear, integro-differential equation that relates the macroscopic motion of a population of particles to a minimal microscopic description of the motion of the individual particles via the partition function. See Eq. 3.5.

Boundary layer. The bottom 1–2 km of the atmosphere in which the frictional effects of the surface are manifest. It is characterized by wind shear with height and turbulent fluxes; it is generally shallower over the oceans.

Catalysis, catalytic agent, chain reaction. A cyclic chemical transformation in which a molecule can facilitate or accelerate the rate of reaction between other molecules while being in its original state at the end of the cycle. Well known stratospheric examples are OH, NO, and Cl: the number of cycles, or chain length of the reaction can be of order 10^4 or more, enabling small amounts of the catalytic agent to destroy large amounts of ozone via the chain reaction [$X + O_3 \rightarrow XO + O_2$; $XO + O \rightarrow X + O_2$]. Other cycles are also effective, for example, R1-R5 in Section 6.2 operates in the polar vortex.

Chaos theory, chaotic. The largely unpredictable, temporal evolution occurring in nonlinear dynamical systems. Despite the system being deterministic, the sensitivity to initial conditions renders the system difficult to predict. Dissipation limits the necessary conservative motion of a point in atmospheric applications.

Chlorine monoxide. A diatomic molecule consisting of one chlorine atom bonded to one oxygen atom. Its unpaired electron qualifies it as a free radical; it is reactive enough to carry, with chlorine atoms, a chain reaction with ozone that is effective in the stratosphere.

Circulation, general circulation. An atmospheric flow pattern with discernible organization that moves air around in a defined volume; general circulation refers to the whole atmosphere. Circulation also has a formal definition involving the integral of the curl of the vector field of velocity (Stokes' Theorem, even though it was originated by Kelvin). The circulation around the boundary C defining the surface S is the product of the area of the surface times the normal component of the curl of the velocity, averaged over S. Given Eq. 3.10. it is intimately bound up with vorticity.

Codimension. If a mathematical object is positioned inside another object of dimension d, it has codimension c if it has dimension d-c. The usage in this book is defined in Eqs. 4.3. and 4.4.

Condensation nuclei. Small particles in the atmosphere, often formed by gas-to-particle conversion, which can grow by accumulating one or more condensable vapours. An example is the formation of sulphuric acid by the oxidation of sulphur dioxide; the low vapour pressure of the acid results in it entering the aerosol phase. Water is, of course, frequently involved in such nucleation processes. A great variety of chemical compositions is known to exist, featuring elements from crustal material, biological and pollution sources.

Convective instability. A parcel of dry air under adiabatic conditions will be buoyant and rise if the temperature gradient dT/dz is steeper (T decreases with height more rapidly) than the dry adiabatic lapse rate (−9·8 K/km). This condition is $d\theta/dz < 0$ if potential temperature is used. For moist air, the release of latent heat has to be considered, potentially making very significant differences below altitudes at which the dry adiabats and the moist adiabats become parallel. See *The Physics of Atmospheres*, J. T. Houghton, Chapter 3, Cambridge University Press, 3rd edition, 2002, for a quantitative treatment.

Coriolis force. Because it is convenient to observe and calculate the atmosphere in the planet's rotating frame rather than in an inertial one, an apparent force f arises on an air element, acting orthogonal to the motion. It is zero when the element is stationary and is directed to the right in the Northern Hemisphere, to the left in the Southern Hemisphere and is equal to $-2\boldsymbol{\Omega} \times \boldsymbol{u}$, where $\boldsymbol{\Omega}$ is the planetary rotation rate. See Eq. 3.17.

Cyclogenesis. Cyclones are characterized by anticlockwise airflow in the Northern Hemisphere and clockwise flow in the Southern Hemisphere. In midlatitudes their development is associated with frontal systems and the attendant weather phenomena. It occurs in baroclinic conditions, with mass divergence (outflow from the cyclone) in the upper troposphere exceeding mass convergence near the surface, to yield a system having low pressure at the centre. Their scales may vary from exceptionally large ones that span the Northern Atlantic to small ones that can be hundreds, rather than the more usual thousands, of kilometres in diameter.

Dimer. The simplest polymer of a stable, single molecule consisting of a pair of the monomers bound in a potential well that is deeper than the value of $k_B T$ at atmospheric temperatures. The water dimer, $(H_2O)_2$, is important in radiative transfer, while the chlorine monoxide dimer, $(ClO)_2$ or Cl_2O_2, is important in polar ozone loss.

Dissipation. The process by which energy in an ordered form is transformed into the least ordered form, which in the atmosphere is the translational motion of air molecules with near-average speed and infrared photons emitted by the atmosphere. The ordered energy enters originally as a beam of high-energy (ultraviolet, visible, and near infrared) solar photons. In terms of mass, a large-scale airflow (for example, a jet stream) may be thought of as ordered; but, it can only exist because an entropy price, through dissipation, has been paid.

Dry adiabatic lapse rate. If an atmospheric column is assumed to be in hydrostatic equilibrium, completely dry and with no energy inputs or outputs, application of the first law of thermodynamics along with the ideal gas law allows calculation of the decrease of temperature, T, with altitude, z: $\mathrm{d}T/\mathrm{d}z = -\Gamma_\mathrm{d}$ where Γ_d is the dry adiabatic lapse rate.

Eddy. A spatial and/or temporal departure of fluid velocity from its average. There are always eddies in the atmosphere no matter over what domains or scales the averages are taken, but it is incorrect to think of them as representing dissipation while the mean represents order.

Electronically excited. A molecular state in which one or more of the electrons is not in the energy ground state. Transitions between such states are often of energies found in the visible and ultraviolet region of the spectrum; in general terms, they also correspond to the amounts of energy needed to break chemical bonds.

Emergent property. A new behaviour that arises in a large enough sample of matter in which conditions are such that enough degrees of freedom interact nonlinearly to sustain properties not evident in smaller or more homogeneous samples. Fluid behaviour emerges in a population of Maxwellian molecules subjected to a flux. Complexity and positive feedbacks are characteristic.

Enstrophy. Enstrophy is defined as half the squared vorticity. The enstrophy equation (Eq. 3.13) can be used to discuss vorticity generation and dissipation under certain assumptions.

Fluctuation-dissipation theorems. This is a general class of relations between the fluctuating force in a system and the magnitude of the dissipation. They are expressed by the Langevin equation (Eq. 3.4) and are fundamental to nonequilibrium statistical mechanics. The first derivations were by Einstein and by Smoluchowski in the consideration of the motion of a particle in a molecular fluid, and used the mean to represent order and the fluctuations (departures from the mean) to represent dissipation. The view from molecular dynamics expressed in Evans and Searles (2002) has fluctuations representing order and near-average molecules representing dissipation.

Fractal. A fractal has structure on all scales, which is in some sense scale invariant, characterized by a fractal dimension, which is not an integer. There appears to be no agreed, rigorous mathematical definition, perhaps not surprisingly, for what can be very complicated objects and behaviours. Regular fractals are generated by repetition of a geometric operation and so are deterministic, whereas natural fractals have randomness and do not repeat exactly, while being statistically similar.

General circulation. See circulation; the broad-scale, average, discernible motion of air that transports mass and chemicals, including water, around in the global atmosphere. Interpretations offered in different coordinate systems can be confusing; the account in D. R. Johnson, *Advances in Geophysics*, Vol. **31**, pp. 43–316 (1989), Academic Press, New York, is recommended.

Generalized scale invariance. A statistical, multifractal approach formulated by Schertzer and Lovejoy (see bibliography and text) and tested against atmospheric observation in both the horizontal and the vertical. It provides for scaling of multiplicative (i.e., nonlinearly interactive) processes in a statistical way, rooted in Lévy stable distributions and using three scaling exponents.

Gravity waves. A solution of the simplified and linearized equations of motion in which internal waves of large vertical and small horizontal scale are considered to propagate vertically with buoyancy being the driving force and gravity the restoring force acting in the context of the static stability. One significant manifestation is to the leeward of steep topography, sometimes called lee or mountain waves. The oscillation is transverse to the direction of propagation, so at least two dimensions must be considered, one of them vertical.

Hurst exponent. H. E. Hurst showed that long-range correlations in the time series of observations of the River Nile extending back to the seventh century A.D. caused persistent, rather than random, power law scaling. If the range $R(t)$ over a span of time t is divided by the standard deviation, $S(t)$, over the same interval, then $R(t)/S(t)$ is proportional to t^H where H is the Hurst exponent; $H = 0.5$ constitutes randomness. $H \sim 0.7$ for the Nile was an unexpected result. See Chapters 27, 37 and 40 of Mandelbrot (1983) for further discussion.

Hydrostatic equation. A simplification of the vertical component of air motion, expressing a balance between the force of gravity and the vertical pressure force. The error in neglecting the vertical components of the earth curvature, friction, vertical acceleration, and the Coriolis force is smaller at larger scales. See Eq. 4.17.

Inertial instability. In a horizontal flow, an air parcel in which the horizontal pressure gradient and the Coriolis force are in steady state, producing wind parallel to the isobars, is said to be in geostrophic balance. If the parcel is displaced horizontally, the displacement will amplify in regions where the wind shear will allow it, resulting in inertial instability—usually found on the equatorward side of a westerly jet stream.

Intermittency. Qualitatively it means that the dissipation of energy in a fluid is inhomogeneous in space and time. The scaling exponent C_1 characterizes the condition in generalized scale invariance. See Eqs. 4.4–4.17.

Isentropic. The existence of the condition of adiabaticity in the atmosphere. In the absence of latent heat transformations, radiative exchange and friction, air moves on isentropic surfaces. These surfaces are actually expressed in meteorology as potential temperature, whose logarithm is proportional to entropy.

Jet streams. A concentrated belt of very fast winds in the upper troposphere and throughout the depth of the periphery of the winter polar stratosphere. The subtropical jet stream is found at an average latitude of 28° in the upper troposphere of both hemispheres, is stronger in winter, and tends to migrate poleward during summer. The polar front jet wanders more in latitude, is less apparent on zonal mean charts, and is associated with midlatitude cyclones. The tropospheric jets are characterized by strong wind shears poleward, equatorward, above and below; cold air is found on the poleward side and warm air on the equatorward side.

Kinetic molecular theory. A theory by means of which the macroscopic properties of gases may be expressed quantitatively, via probabilistic analysis of the motion of the constituent molecules, through either classical or quantum mechanics.

Laminar flow. An idealized state in which all the constituent elements of a fluid possess the same nonzero velocity. Observations of wind shear and consideration of the dynamics of a molecular population under a flux make it clear that it is an unattainable condition in the atmosphere, on any scale.

Lapse rate. An expression for the change of temperature with altitude. The dry adiabatic lapse rate, Γ, is defined as $\mathrm{d}T/\mathrm{d}z = -g/C_p = -\Gamma$ and care is needed in observing the sign convention. Γ has the value $9.8\,\mathrm{K\,km^{-1}}$ for the dry atmosphere; in the more real moist atmosphere, allowance for the phase transitions of water yields an average value in the lower troposphere of $6.5\,\mathrm{K\,km^{-1}}$.

Law of mass action. In a homogeneous system the rate of a chemical reaction is proportional to the active masses of the reacting substances. Active mass is usually taken as molecular concentration. At equilibrium, it leads to the expression of the equilibrium constant as the ratio of the rate coefficients of the forward and back reactions.

Lévy exponent, -index, -stable distribution. A Lévy stable distribution is a class characterized by four exponents (see Section 7.1) the stability index, α, the scale factor σ, the skewness β, and the mean μ. α and β determine the shape of the distribution, while σ and μ determine scale and shift. There is a generalized central limit theorem in which these distributions play an important role.

Long tail. Also referred to as 'fat tail'. A colloquial description in which there is larger population at values far removed from the most probable, relative to a Gaussian. It is often a sign of power law behaviour, signalling the presence of long-range correlations, frequently produced by positive feedback.

Lorentzian lineshape. The theoretical shape of a spectral line in conditions where it is determined by pressure broadening; that is, it is dominated by the effects of intermolecular collisions upon the energy of the quantum levels involved in the optical transition. A key assumption is that collisions are strong enough that no memory of phase or velocity endures in the next collision. Departures from this shape are observable in certain atmospheric lines. See Goody and Yung (1989), Section 3.3.

Maximum entropy, -production, -principle. In the event of incomplete information about a system being available, the maximum entropy principle is a technique that allows selection of a probability distribution that is as unbiased as possible, given the known constraints. It maximizes (a) ignorance, (b) uncertainty, (c) randomness, and (d) lack of bias, depending upon nomenclature. It aims to ensure exclusion of unavailable information. Maximum entropy production applies to a fluctuating, steady-state system held in a nonequilibrium state by application of an external flux of energy: the system responds by maximizing the rate at which entropy is produced.

Maxwellian molecules. Hard, structureless, spherical particles undergoing elastic collisions with each other ('billiard balls').

Meridional circulation. A 'flow' of air in the latitude-height plane produced by averaging around circles of latitude. The circulation is a statistical construct rather than being a physical representation of what air actually does, which pathologically involves all three dimensions.

Molecular dynamics. The application of numerical methods via digital computers to calculate the behaviour of molecular populations in terms of the movement of individual molecules. See Alder (2002) and Rapaport (2004).

MOZAIC. An acronym for Measurements of OZone and water vapour by in-service Airbus airCraft. A programme that since 1993 has collected observations from five A340 commercial airliners in service on worldwide routes. See Marenco et al. (1998).

Multifractal. The scaling behaviour of a monofractal is describable by one variable, its fractal (noninteger) dimension. The fractional dimension can be exemplified by the extent to which a one-dimensional line can fill a plane (two-dimensional space); it will be more than 1 and less than 2. A multifractal has value at every point and requires more than one exponent to describe it. Such objects are generated by multiplicative interaction of independent, random processes. In generalized scale invariance, three exponents are required: H_1, the conservation exponent; C_1, the intermittency, and α, the Lévy exponent.

Navier–Stokes equation. The set of differential equations that describes the motion of a fluid based on the conservation of mass, momentum, angular momentum, and energy, in addition to the equation of state for the fluid (the constitutive relation). They are rarely discussed or stated in their full form, nor are the limits of the continuum assumption fully explored. See, however, http://www.navier-stokes.net/nscont.htm, where M. S. Cramer provides an interesting discussion.

Nitrous oxide. The linear molecule $N = N = O$ that is, via reaction with $O(^1D)$ atoms, the principal source of odd nitrogen ($N + NO + NO_2 + NO_3 + 2N_2O_5 + HNO_2 + HNO_3 + HNO_4$) in the middle and upper stratosphere. It is a conservative tracer (passive scalar) in the lower stratosphere and has been widely used as such in both general circulation models and in observational studies. It is produced biologically and has an atmospheric residence time of $\sim$115 years.

Numerical model. A suite of programs run on digital computers that seek to represent the atmosphere and its boundaries as an initial value problem in mathematical physics by integrating the relevant differential equations forward in time.

Ozone. A triatomic gas produced by the recombination of free oxygen atoms with oxygen molecules in a three-body process following production of the former upon photodissociation of the latter by solar photons with wavelengths <242 nm. It has absorption bands ranging from the ultraviolet to the near infrared: Hartley, Huggins, Chappuis, and Wulf.

Parametrization, -subgrid scale. A formulation in computer models that represents dissipation on scales smaller than those resolved by the model grid, currently about 50 km in global weather forecasting models and 300 km in global climate prediction models, in the horizontal. Smaller grids, down to 1–6 km, are used in local forecasting models. Without it, aliasing and energy accumulation occur on the smallest resolved scales.

Passive scalar. A term used in turbulence theory to describe what in meteorology and atmospheric chemistry is called a conserved tracer.

Persistence. A condition that strictly applies to data in which the H_1 exponent suffices; that is to say, there is no intermittency. There is neighbour-to-neighbour correlation on all scales and $H_1 > 0.5$. $H_1 = 1$ corresponds to complete correlation, a smooth curve with perfect persistence.

Photochemical kinetics. Sequences of reactions in a gas initiated by absorption of a photon followed by a series of elementary collisional steps involving one or more reactive species resulting from the photon's energy. Reactions R1-R5 in Section 6.2 are atmospheric examples occurring in the stratospheric winter polar vortex.

Photodissociation. The act of a molecule splitting into photofragments following the absorption of a photon sufficiently energetic to break one or more of the chemical bonds in the parent molecule. It is the process which drives virtually all atmospheric chemistry.

Photoisomerization. A process when a molecule rearranges into a different atomic configuration with the same molecular weight following the absorption of a photon. An example is the absorption of a photon by OClO resulting in a rearrangement to ClOO (Vaida et al. 1989).

POLARIS. The acronym for Polar Ozone Loss in Arctic Regions in Summer; an airborne mission using the NASA ER-2 based in Fairbanks, Alaska, at intervals between April and September 1997. See Newman et al. (1999).

Potential enstrophy. The average of half the squared potential vorticity. Often used to characterize cascades across scales in idealized geophysical flows. See Pedlosky (1987), Chapters 3 and 4.

Potential temperature. The potential temperature, θ, is defined by Poisson's equation, $\theta = T(1000/p)^{R/C_p}$, where T is absolute temperature in degrees Kelvin, p is pressure in millibars (hPa), R is the gas constant, and C_p the specific heat of air at constant pressure. See Eqs. 3.17–3.19 for an example of usage. It is the temperature an air parcel would have were it to be moved isentropically to 1000 hPa pressure, that is, to the surface; it is meteorology's way of accommodating the third law of thermodynamics. Potential temperatures in the stratosphere range from about 340 to 2000 K; it is thus apparent why it is rare to find unmixed stratospheric air at the surface!

Potential vorticity. See Eq. 3.19. In addition to being a tracer for atmospheric motion, calculable from the standard meteorological variables, it is involved in relations determining large-scale atmospheric flow in the limit of no dissipation. See Eq. 3.19; also Hoskins et al. (1985) and Chapter 2.5 of Pedlosky (1987).

Pressure tendency equation. Usually discussed in the context of subjective forecasting techniques based upon simplifications and linearizations of the governing equations. It relates the rate of change of surface pressure at a point to the mass convergence in the overlying air column; it is relevant because vertical motion strongly influences weather, with ascent associated with clouds and precipitation and descent with fair weather. Friction, orography, and diabatic heating all affect the surface pressure tendency in the actual atmosphere. In practice, the compensation between convergence and divergence can be significantly local in the altitude of a total air column, thus limiting the usefulness of the equation at the surface.

Radiosonde. A package of instruments to measure winds, pressure, temperature, and humidity, launched at a few thousand stations around the globe, usually at 00 UTC and 12 UTC with a lesser number at 06 UTC and 18 UTC. The balloons reach altitudes up to 30 km, ascending at $\sim 5\,\mathrm{m\,s^{-1}}$ and telemetering their observations back to the launch site.

Rayleigh-Taylor instability. This instability occurs when a denser fluid floats above a less dense fluid; any perturbation of the boundary will amplify, as contrasting fluid elements seek to exchange places in the gravitational field. Atmospheric examples are the 'mushroom clouds' from nuclear explosions and volcanic eruptions.

Reactive nitrogen. Species such as N, NO, NO_2, NO_3, $2N_2O_5$, HNO_2, HNO_3, HNO_4. The phrase is often used to describe the sum of the nitrogen atom concentrations bound up in these molecules. Molecular nitrogen, N_2, and nitrous oxide, N_2O, are excluded and so usually are nitrogenous organic molecules.

Ring currents. The vortex-like structures formed on very short time and space scales when an equilibrated population of Maxwellian molecules is subjected to an anisotropic flux. See Alder and Wainwright (1970). Their importance lies in the fact that fluid behaviour has emerged from a molecular population.

Rovibronic. A compound adjective describing the simultaneous occurrence of changes in rotational, vibrational, and electronic quantum numbers in an optical transition between molecular energy states.

Scale invariance. An object with structure over a range of scales whose appearance is independent of the resolution at which it is viewed. This can be statistical and can involve more than one scaling ratio.

Scaling. An object is said to exhibit scaling behaviour when it has no characteristic length scale and when it obeys power law statistics.

SOLVE. SAGE-III Ozone Loss and Validation Experiment. SAGE stands for Stratospheric Aerosol and Gas Experiment, a solar occultation satellite in an inclined orbit. SOLVE took place between December 1999 and March 2000, using the NASA ER-2 and DC-8 aircraft based in Kiruna, Sweden. See Newman et al. (2002).

Stratosphere. The 10–15% by mass of the atmosphere having its lower boundary as the tropopause at 9–17 km altitude and its upper boundary as the stratopause at 45–50 km. It is characterized by a large-scale temperature inversion caused by the heating due the absorption of solar radiation by ozone, rendering it non-convective and statically stable.

Stratospheric polar vortex.The strong cyclonic circulation (westerly winds) around the winter pole, caused by the dynamical response to wintertime radiative cooling at the pole, combined with the reservoir of vorticity and potential vorticity at high latitudes arising from the planetary rotation.

Structure function. See Eq. 4.8. The slope of plots of the logarithm of the structure function against the logarithm of the parametric interval (lag), if linear, provides an estimate of the scaling exponent. Higher moments are very demanding, requiring large volumes of data.

Statistical multifractals. Regular fractals involve a repeated geometric operation, such as a Koch snowflake, and are deterministic. Most natural occurrences involve both multiplicative processes and randomness and never repeat exactly; they require statistical description, hence the name.

Synoptic scale weather. The alternating sequence of cyclones (low surface pressure at the centre) and anticyclones (high surface pressure at the centre) causes the synoptic scale weather; each particular system will typically have a scale of a few thousand kilometres, but they vary considerably in size on an individual basis.

Transport. The transfer of air, including its content of trace species, from one location to another in the atmosphere. It cannot be estimated accurately using Lagrangian techniques alone, advection alone, or diffusion alone. It inescapably involves all three dimensions, is scale invariant and involves both advective and mixing processes.

Tropopause, -folding. The World Meteorological Organization definition is that the tropopause is at the altitude above which the lapse rate decreases to less than $2\,\mathrm{K\,km^{-1}}$ throughout a layer at least 2 km deep. It is sometimes called the 'thermal tropopause', to distinguish it from the 'dynamical tropopause', which is in the $1\text{–}3 \times 10^{-5}\,\mathrm{Km^2\,kg^{-1}\,s^{-1}}$ range in potential vorticity, the defining variable. There are also 'cold point' tropopauses used in the tropics and 'ozone' tropopauses selected on the basis of ozone mixing ratios in the 50–100 ppbv range (Bethan et al. 1996). All have limitations; the thermal tropopause is usually easy to assign on a thermodynamic diagram, using routine meteorological observations. The ambiguities that appear in the definitions from observations signify the importance of exchange across the tropopause, induced in midlatitudes by the folding of the potential vorticity contours near jet streams. There is very often the signature of some fraction of stratospheric air in the upper troposphere and of tropospheric air in the lower stratosphere, irrespective of location. Multiple tropopauses have long been observed near the subtropical jet stream, when mid-latitude and subtropical tropopauses are geographically coincident and the air above the former and below the latter is intermediate in character between stratospheric and tropospheric.

Troposphere. The 85–90% of the atmosphere found between the surface and the tropopause and which is subject to overturning by convective processes.

Variogram. See Section 4.2. It is a measure of variance in a dataset as a function of distance or time and, therefore, describes spatial or temporal correlation.

Vorticity. It is the curl of velocity; see Eq. 3.10. Conventionally positive in the Northern Hemisphere for anticlockwise, that is, cyclonic rotation; if it is positive there, the local parcel possessing it is rotating about its own vertical axis as it moves with the flow. The vector is directed upwards along this axis. High values are correlated with upward air motion and development of low pressure at the surface.

WAM. An acronym for WB57F Aerosol Mission, which took place from NASA Johnson Space Center, Ellington Field, Houston, April–May 1998. Via the Particle Analysis by Laser Mass Spectrometry (PALMS) instrument, the first real time analysis of the chemical content of individual aerosol particles was achieved in the upper troposphere and lower stratosphere.

Westerly. A wind that blows from the west. The average wind in the mid-latitude troposphere is westerly; that is to say, the air moves faster in the same direction that the solid planet rotates.

Index

Note: page numbers in *italics* refer to Figures and Tables, whilst those in **bold** refer to Glossary entries.